WITHDRAWN

A Global View
of Energy

A Global View of Energy

Edited by
Behram N. Kursunoglu
Center for Theoretical Studies,
    University of Miami
Andrew C. Millunzi
U.S. Department of Energy
Arnold Perlmutter
Center for Theoretical Studies,
    University of Miami

Associate Editor
Linda Scott
Center for Theoretical Studies,
    University of Miami

LexingtonBooks
D.C. Heath and Company
Lexington, Massachusetts
Toronto

Library of Congress Cataloging in Publication Data

Main entry under title:
  A Global view of energy.

  1. Power resources—Addresses, essays, lectures. 2. Energy policy—Addresses, essays, lectures. I. Kursunoglu, Behram, 1922-   . II. Millunzi, Andrew C. III. Perlmutter, Arnold, 1928-   .
TJ163.24.G58   333.79   81-47525
ISBN 0-669-04647-7   AACR2

Copyright © 1982 by D.C. Heath and Company

All rights reserved. No part of this publication may be reproduced or transmitted in any form or by any means, electronic or mechanical, including photocopy, recording, or any information storage or retrieval system, without permission in writing from the publisher.

Published simultaneously in Canada

Printed in the United States of America

International Standard Book Number: 0-669-04647-7

Library of Congress Catalog Card Number: 81-47525

Contents

	Foreword *Karl Cohen*	ix
	Acknowledgments	xvii
Chapter 1	French Energy Policy for the 1980s *Jean Claude Balaceanu*	1
Chapter 2	The Urgent Need for Realistic Planning Now *Mike McCormack*	19
Chapter 3	Getting from Here to There: The Energy Road Ahead *M.E.J. O'Loughlin*	27
Chapter 4	Geopolitics of Energy in the Persian Gulf *Edward Teller*	39
Chapter 5	Annotation on Energy in the Crucial Decade of the 1980s *C. Pierre L. Zaleski*	43
Chapter 6	Toward a New Order for the Global Energy System *Wolfgang Sassin*	47
Chapter 7	International Aspects of U.S. Energy Choices *Sam H. Schurr*	59
Chapter 8	Energy in the Developing Countries *Richard H. Sheehan*	67
Chapter 9	The Role of Coal *Leslie Grainger*	79
Chapter 10	The Role of Electricity as a Substitute for Liquid Fuels *L.G. Hauser*	89
Chapter 11	The Disequilibrium Effects of Oil-Price Shocks *Narasimhan P. Kannan* and *Martin M. Scholl*	103
Chapter 12	The Role of Electricity in Solving Energy Problems in the Near Term *Thomas H. Lee*	113

Chapter 13	Relationship between Energy Growth and Economic Growth: The Point of View of the Developing Countries *Marcelo Alonso*	119
Chapter 14	Alternative Energy Options for Puerto Rico *Eduardo López-Ballori*	133
Chapter 15	Near-Term Feasibility of Candor Fusion Reactors *Bruno Coppi*	141
Chapter 16	Inertial Fusion and Energy Production *John F. Holzrichter*	151
Chapter 17	Fusion and U.S. Energy Policy *N. Douglas Pewitt*	159
Chapter 18	Status of U.S. Tokamak Effort *Paul J. Reardon*	165
Chapter 19	Nuclear Fission as a Global Energy Resource *Robert J. Creagan*	201
Chapter 20	Nuclear Energy—The Next Phase: Approaches to Restoration of Confidence and Motivation *Peter Fortescue*	217
Chapter 21	How Available Is the Nuclear Option? *William H. Hannum*	225
Chapter 22	Comparisons of the Health Effects of Energy Systems: An Assessment for France *Philippe Hubert*	233
Chapter 23	Health Hazards Associated with Electric-Power Production: A Comparative Study *W. Paskievici*	249
Chapter 24	Environmental Risks of Energy Production: The Carbon Dioxide Example *Robert S. Chen*	275
Chapter 25	Risk-Assessment Analysis for Various Energy Sources: Some Annotations *Klaus Gottstein*	301

Appendix: Program	309
List of Contributors	317
About the Editors	319

Foreword: Combined Findings of Four International Scientific Forums on Energy

The fourth of a series of interdisciplinary international scientific forums, attended by scientists from sixteen countries and three international organizations, was held in Fort Lauderdale, Florida, on 10-14 November 1980. The first forum, on the subject "An Acceptable Nuclear Energy Future of the World," was held in Fort Lauderdale on 7-11 November 1977. The second forum extended the horizon from nuclear energy to all energy sources. Its title was "An Acceptable World Energy Future" (held in Miami Beach, 27 November to 1 December 1978). The third forum, entitled "Energy for Developed and Developing Countries," was held in Nice, France, from 29 October to 2 November 1979. It focused specific attention on the energy problems of the developing countries, since it had become clear in the earlier forums that these problems were particularly serious. The fourth forum on "Geopolitics of Energy" further pursued the issue of the interactions between the energy problems in developed and developing countries.

Following the first three forums, energy-policy findings and recommendations were published, based on material presented and discussed during the sessions. The findings were formulated by the planning committee in areas in which substantial (if not necessarily unanimous) consensus had been reached. The findings and recommendations of the previous forums were reviewed by the planning committee in the light of material presented in the fourth forum. A combined set of findings was developed and divided into the following categories:

1. previous findings and recommendations that were confirmed during the fourth forum;
2. new findings based on discussions unique to the fourth forum;
3. previous findings that were modified as the result of information discussed in the fourth forum;
4. previous findings that were not reviewed during the fourth forum.

Category 1: Previous Findings Confirmed

Energy Demand

In spite of higher energy prices and considerable improvement in the efficiency of energy use, energy demand will continue to grow, though not at the rates of the last twenty years. World demand is likely to have doubled early in the next century. This demand estimate is contingent on a much needed economic improvement in the developing world. Failure to meet this demand will result in extensive social ills, such as poverty, starvation, unrest, epidemics, riots, and wars.

No single technology can meet the world's future demand. It is likely that all technologies—conventional fossil fuels, nuclear fission, nuclear fusion, geothermal, and solar—will be required to meet the qualitative and quantitative aspects of demand, just as today no single technology meets all demands. Meeting the energy demand of the still rapidly increasing world population with legitimate expectations of a higher standard of living calls for large-scale mobilization of labor, materials, capital, and technical and management skills. It should be a constant preoccupation to accomplish this economically and effectively to avoid overtaxing the world's resources of these necessities.

Energy Modeling

Energy models project alternative futures of energy demand, supply, and price movement. Differences between projections and between models reflect intrinsic uncertainties about future economic behavior and about the success and cost of future technologies. There is, however, a growing consensus that changes in energy demand or modes of energy supply occur slowly. Public policy must recognize this characteristic. To be effective, a public energy policy must be timely and consistent over long periods. Further, in view of the uncertainties mentioned, public policy should be based on a range of estimates that reflects different contingencies.

Urgency of World Energy Problem

There is an urgency to the world energy problem that, especially in view of the long lead times, brooks no delay in determining and executing national programs and in seeking international cooperation to take up the tasks and share the benefits equitably.

Nuclear Fission

Nuclear fission must play a significant role in meeting world energy demand over at least the next five decades, and over this period it cannot be forgone without excessive risk. An assured nuclear-fuel supply, of utmost importance

Foreword

to many nations, cannot be guaranteed by uranium mining alone. Although the urgency will vary from country to country, fuel reprocessing is essential in the application of nuclear-fission energy. Further, the best way to handle spent fuel and take care of nuclear wastes is to reprocess the spent fuel.

There are many candidate systems that may be called on to supplement or, eventually, replace our present largely light-water-reactor (LWR) technology. These include fast breeder reactors, high-temperature gas reactors, heavy-water reactors, and homogeneous reactors. Developments in all these systems should be pursued vigorously on an international basis, though not necessarily all systems in all countries. Practical consideration of the ability to produce and deploy reactors in the numbers necessary dictates that currently successful systems be sustained and their installation encouraged by government until advanced systems are fully available and acceptable technically, economically, and industrially.

The plutonium/uranium fuel cycle has particular advantages in fast-spectrum reactors, and the uranium-233/thorium fuel cycle has advantages in thermal reactors. Both will need to be developed, including all necessary steps for full implementation.

Progress in Nuclear Fusion

Impressive progress has been achieved toward proving the scientific feasibility of fusion systems based on the principles of magnetic and inertial confinement. Progress has been made also that suggests systems that, on a longer time scale, may indicate economic feasibility. Development of these systems, already involving a considerable degree of international cooperation, should be pursued vigorously on this basis—again, not all systems in all countries. However, the possible successful development of fusion technology should not delay the prudent and necessary deployment of fission technology. It is possible that the first application of fusion technology will be in a hybrid fission/fusion system.

Financing Energy Investments

Despite recent price increases and those foreseeable in the future, investments and expenditures for energy in both developed and developing countries over the next two decades do not appear excessive compared with national incomes and historical investment patterns. Financing of energy investments should be available for countries with adequate energy planning and stable policies based on these plans.

Conservation of Energy in Developed Countries

Conservation measures and an increase in domestic production of energy in developed countries, besides being beneficial in themselves, have the additional advantage of increasing the availability of oil for developing countries.

Public Safety: Risks and Benefits

Standards for environmental protection—for example, against carcinogens in foods or air pollution from plant emissions—should be stable, rational, and applied across the board to all risks. Ad hoc application of standards to some hazards and neglect of others result in unbalanced application of resources and unjustifiable public anxieties.

Atmospheric Carbon Dioxide

The possibility of serious climatic changes from the accumulation of anthropogenic CO_2 in the atmosphere, as from burning of fossil fuels and deforestation, is widely recognized. However, our knowledge of both the global CO_2 balance and the global climatic effects of CO_2 accumulation is rudimentary. During the next decades a sustained international scientific study, incorporating data gathering and theoretical analysis, should be implemented. Regulations against the use of fossil fuels, in advance of fuller understanding of the phenomena and of possible countermeasures, are premature.

Category 2: New Findings

Vulnerability of World Oil Supplies

Forty percent of the oil supply of the free world passes through the Persian Gulf. The prolonged interruption of these supplies by any one of a number of scenarios (for example, Soviet expansionism, revolution, or religious strife) is a real possibility. If this occurred, the economies of the industrial democracies and of the developing nations would be in jeopardy. Military action is a conceivable response to such a situation, but the results could be catastrophic. Immediate programs to reduce this vulnerability should be instituted. International consultation and joint planning are required.

Foreword

Human Capital in Developing Nations

The infusion of outside capital and technology into developing nations, though necessary, is not sufficient for their economic progress. Another necessary ingredient is an increase in human capital through education, motivation, and leadership development.

Category 3: Previous Findings Modified

Public Acceptance

The forum once again considered the increasing polarization and hardening of public opinion for and against particular energy alternatives. This state of affairs points up the need for continuing efforts to provide accurate and impartial information. There is a tendency in some countries to resort to increasingly labyrinthine trial-type procedures, sometimes before several agencies, to resolve the social and political issues involved in energy decisions. Although the growth of such complicated and duplicative procedures is intended to compensate for both a basic fear of complex technology and a lack of confidence in our institutions to govern properly, such procedures have enormous costs for society in delay, uncertainty, and in some instances frustration of public policy without any demonstrable benefit over alternative processes. Various forms of less complex decision making have produced not only excellent technical analyses but also final decisions on a timely basis and with wide public acceptance—for example, the Windscale proceedings in Great Britain. Therefore, these less ritualistic modes of decision making, sometimes styled "legislative," are recommended as being more appropriate to governmental energy decisions.

The positive example of France and the negative examples of Sweden and Austria show that a firm governmental nuclear policy is a major component of public acceptance of nuclear power.

Nuclear-Reactor Accidents

The probability that accidents in existing reactors will cause harm is acceptably small and, we believe, with proper use of experience will diminish even as the number of reactors increases. A nonadversary relationship between government and the nuclear industry, a joint commitment to accident prevention, and the establishment of logical numerical safety goals will improve both nuclear-reactor safety and nuclear-reactor economics.

Category 4: Previous Findings Not Reviewed

Regulation of Energy Demand

Although it may be necessary to correct market prices of energy to take account of other factors, such as social costs, in general regulation of energy demand by price rather than by external constraints, such as allocations, is preferable.

*Technical Cooperation between Developed
and Developing Countries*

Of the many constructive measures of cooperation between developed and developing countries currently used, particular emphasis should be placed on joint activities (training, research and development, practical applications) carried on within developing countries by mixed teams. International research and development in energy technologies of particular interest to developing countries, such as biomass and biological engineering, are highly desirable.

Biomass and Solar Energy in Developing Countries

Biomass and solar energy are already important elements of the energy budget in many developing countries. However, it is unlikely that a large fraction of their future energy needs can be satisfied in this way.

Solar Energy

Although several solar-energy systems are competitive today in certain regions of developed countries, their costs will in general be high in the foreseeable future compared with those of more conventional energy sources. Yet solar, wind, and biomass are expected to supply a significant fraction (excluding hydropower, of the order of 5 percent) of the energy needs of developed countries by the year 2000.

Oil and Gas

Considerable oil and gas resources exist that can be developed into reserves at prices between today's OPEC prices and the probable much higher prices

Foreword

of synthetic oil and gas from coal. It is not entirely clear whether recent OPEC price increases reflect a real scarcity value, according to well-known economic theory, or are politically inspired.

North American Region

A freer exchange of technologies and resources between Canada, Mexico, and the United States would be mutually advantageous.

Energy and Economic Growth

Although energy use is pervasive throughout industrialized economics, substantial reductions in the growth of energy consumption are possible with only relatively small reductions in the growth of economic output (GNP). The precise GNP-growth reductions associated with an energy-growth restriction depend on the elasticity of energy demand, the value share of energy in the economy, the economic policies undertaken, changes in the rate of capital formation, and whether the reductions are motivated by cost increases or by other policy measures.

Conservation

Largely because of the success of OPEC, energy demand has grown less than it otherwise would have and has been paid for at higher prices. Therefore, more efficient use of energy in many instances has become cost effective. The time scale for introduction of these improvements is usually long, being related to the turnover time of huge existing stocks of buildings, transportation equipment, and industrial-process equipment. Consistent long-range public policies will help accelerate the reduction of energy-demand growth through more efficient use of energy. Such policies should include an R&D strategy for improved energy use commensurate with the expected high future prices of new energy supplies.

Standardization

Standardization of complex industrial equipment is highly desirable. Reliability, safety, and economics are all improved by the learning process. Changes in standards must be rationed carefully to avoid unexpectedly negative effects because of interruption of the learning process. Plant-unique requirements imposed by local governments should be avoided for the same reason. Scaling up plant sizes to obtain economies of scale must also be done circumspectly.

Proliferation of Nuclear Weapons

It is recognized that the deployment of fission power or hybrid fusion/fission power on a large scale poses problems of safeguards of material against proliferation of nuclear weapons. We are confident that political, institutional, and technical measures can be taken by the international community that will be effective in diminishing the risk of proliferation while retaining the economic advantages of nuclear power. Therefore, we do not believe that the risk of proliferation should deter the use of nuclear energy.

Reprocessing and breeder-reactor development are economically unattractive for small-scale nuclear-energy systems. Widespread economically motivated use of these technologies is a distant prospect. On the other hand, policies designed to prevent the spread of nuclear weapons that interfere with a nation's legitimate economic goals will not be successful. Multilateral accords between weapons and nonweapons states are necessary to manage this problem. Unilateral actions by weapons states will be counterproductive.

Karl Cohen
Forum Rapporteur

Acknowledgments

This book grew out of the fourth in a series of international scientific forums on energy, which focused its deliberations on geopolitics of energy. It is a special pleasure to extend our gratitude to session moderators, dissertators, and annotators for their work, which made it possible to organize a good forum on an important topic. The fundamental conclusions of the fourth forum, as seen by our forum rapporteur, Dr. Karl Cohen, are discussed in the foreword. We would like to take this opportunity to record our grateful appreciation to the following corporations for their support of the Center for Theoretical Studies in organizing this series of forums: Exxon Education Foundation, Mobil Research and Development Corporation, Westinghouse Electric Corporation, The Chase Manhattan Bank, N.A., Alcoa Foundation, Dresser Industries, Allied Chemical Company, General Electric Company, Union Carbide Corporation, and Grumman Aerospace Corporation.

A Global View of Energy

French Energy Policy for the 1980s

Jean Claude Balaceanu

Introduction

France is poorly endowed with energy resources and, since the 1973/1974 oil crisis, has faced a transformation of world energy resources that is unprecedented in history (see figure 1-1). It is not surprising that full awareness of the energy crisis has come more quickly and keenly in France than in other countries for at least three primary reasons. The first of these is the trend in prices. In the space of just seven years, the price of oil has increased more than sevenfold in constant dollars. The second factor is the trend in production. Actually, there is no overall shortage of energy reserves on the planet, not even of oil. But there is a problem of flow. Figure 1-2 shows possible scenarios for oil supplies. Whatever happens, the oil market will remain extremely tight if the consumer countries do not make a sizable effort to reduce their demand.

Finally, there is, France's vulnerability. France is 75-percent dependent on imported energy (see figure 1-3) and therefore faces the coming decade under difficult conditions. Above all, France will have to count on its own efforts to loosen the noose of energy constraints and to reduce its dependence on oil (see table 1-1).

France's Energy Policy for 1990

On 27 March 1980 the Conseil de Planification (Planning Council) defined the main objectives of the French energy plan for 1990. There are three main guidelines for this policy (see figure 1-4):

1. to promote energy conservation, with the emphasis on using less oil;
2. to make greater use of forms of energy that can be substituted for oil, such as nuclear energy in particular, along with coal and renewable energy resources;
3. to reduce France's oil vulnerability by prospecting for hydrocarbons at home, mastering oil technology, and improving relations with the producing countries.

1

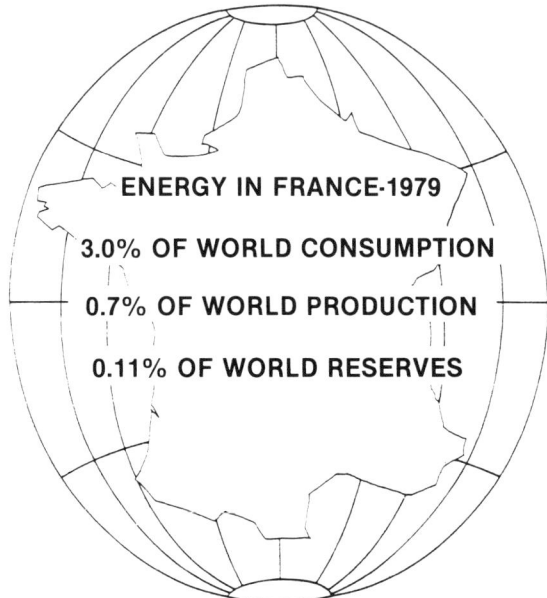

Figure 1-1. Energy in France, 1979

French energy production will more than double over the coming decade (109.5 million tons of oil equivalent (MtOE) in 1990 compared with 48.5 MtOE in 1979) thanks to the development of the electronuclear program and the promotion of new energy sources (see table 1-2). The gap between French consumption and production of energy will have greatly narrowed. France will produce 45 percent of the energy it consumes as opposed to 25 percent in 1979. Likewise, it will have reduced its energy-dependency rate from 75 to 55 percent. At the same time, nuclear energy will supply a larger share of the energy consumed than will oil (73 MtOE against 68 MtOE; see figure 1-5).

French energy consumption will grow from 3.87 million barrels per day (Mbbl/d) in 1979 to 4.84 Mbbl/d in 1990, a very moderate increase (about 2.1 percent per year, assuming a 3.5-percent average increase in the gross national product (GNP) during the period) owing to vigorous energy-saving efforts (see figure 1-6). The energy consumption pattern will be very different from what it is now; the share of nuclear energy will increase sevenfold, from 4.5 to 30 percent. The share taken by coal and gas will remain stable, and that taken by renewable energy sources will rise as the result of a rapid increase in the use of new energy sources. The share of oil will be reduced from 56 to 30 percent (see figure 1-7).

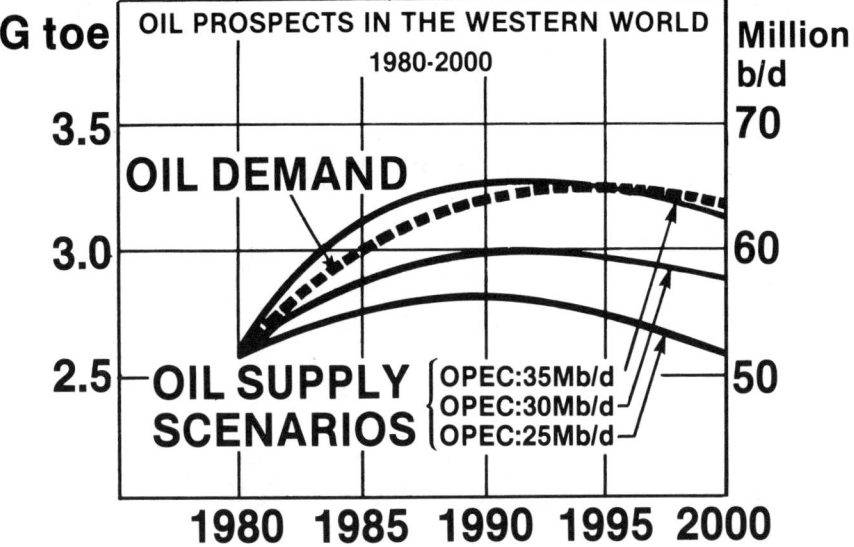

Figure 1-2. Oil Prospects in the Western World

The Pursuit of Energy Conservation

The conservation effort is especially difficult in that by 1973 France was already a particularly economical country with respect to energy consumption (see figure 1-8). Energy savings in 1990 will be approximately 60 million tons (Mt) of oil compared with 18 Mt in 1979, a threefold increase in the space of a single decade (see figure 1-9). To reach this goal, the growth of energy consumption will have to be limited to a rate 40 percent lower than the growth rate of the gross domestic product. In other words, for a GNP rising at an average annual rate of 3.5 percent, energy consumption will rise by only 2.1 percent.

The rest of this section deals with the applications of France's 1990 energy-conservation policy in the different sectors (see figure 1-10 and table 1-3).

Industry

Industry will experience a large-scale redeployment of energy, with the aim of reducing oil consumption by a factor of 2.7. To do this, it will be necessary to: (1) change two out of three boilers; (2) increase coal consumption fivefold; and (3) double the rate of conversion to electricity in industry

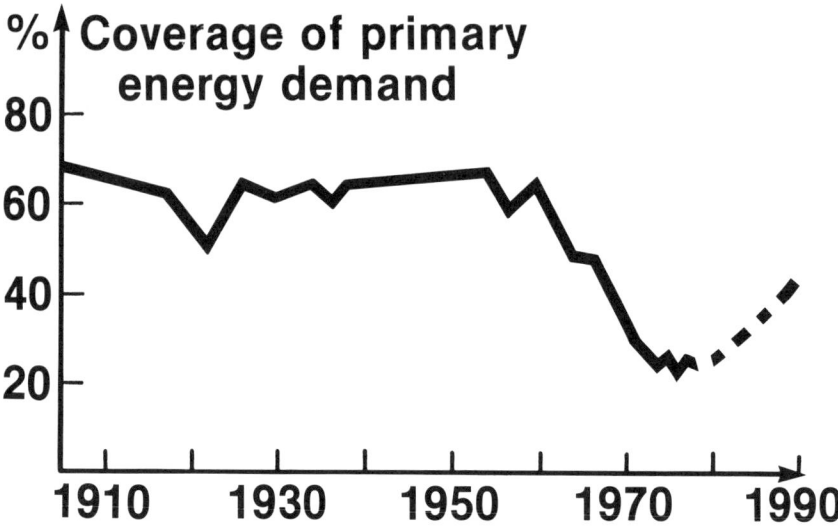

Figure 1-3. The Extent of French Energy Independence

in comparison with the rate of the last few years. A different consumption pattern will appear, as illustrated in figure 1-11. Oil's share will drop sharply, from 38.5 percent in 1980 to 10.5 percent in 1990. At the same time the share of electricity will rise by about 7 percent and that of gas by 8 percent. But the most spectacular trend will lie in the sharp increase in coal consumption, which will jump from 3.5 percent of the total in 1980 to 17 percent in 1990.

The Residential and Tertiary Sectors

Particular efforts will be directed toward the residential and tertiary sectors. For example, a 30-percent reinforcement of the insulation of new homes

Table 1-1
A Few Significant Changes from 1973 to 1979

GNP	+20%
Cars	+30%
Houses and flats (with central heating)	+40%
Washing machines	+35%
Energy consumption	+8.8%
Oil consumption	−7%

French Energy Policy for the 1980s

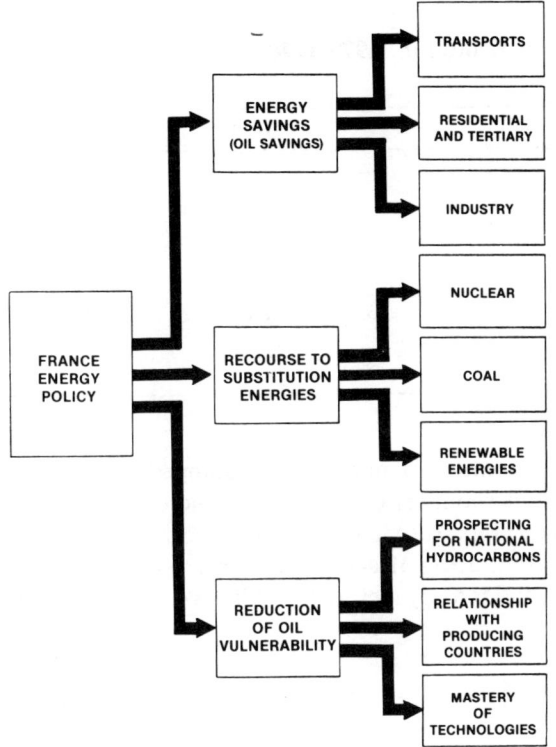

Figure 1-4. World Energy Demand: Changing Perceptions

and offices will provide savings of 6 MtOE. At the same time, the insulating of 500,000 existing homes per year is planned, notably when they are being converted to gas or electric heating. Parallel efforts will be made for offices and shops, resulting in savings of 2 MtOE. All in all, the energy consumption of a new house will be cut in half by 1985 (see figure 1-12).

The residential and tertiary sectors represent an especially important area for the development of new forms of energy. The goal for 1990 is to use new forms of energy in 5 million homes, as follows: (1) 1.5-2 million homes heated by wood; (2) 2 million homes equipped with solar water heaters and 1.5 million homes using solar energy for space heating (corresponding to savings of 15-20 percent); and (3) 800,000 homes using geothermal heating.

As for electricity, the rate of conversion to heating by electricity should be aligned with that of progress in the field of electronuclear energy. With this in mind, 2 million homes will be equipped with electric heating by 1985.

Table 1-2
Energy-Production Evolution, 1979-1990
(Mtoe)

Primary Energies	1979	1990
Nuclear	8.3	73
Hydro	14.7	14
Coal	13.3	
Natural gas	7.1	} 12.5
Oil	2.1	
New energies	3	10
Total	48.5	109.5
Coverage of primary energy demand	25.1%	45.2%

Subsequently, the number of new homes equipped with heating systems running on electricity should rise at a markedly increased rate. Likewise, the use of gas heating will increase in other new homes. Given the combined effect of these different areas of action, the pattern of energy consumption for the residential and tertiary sectors (including water heating) will undergo

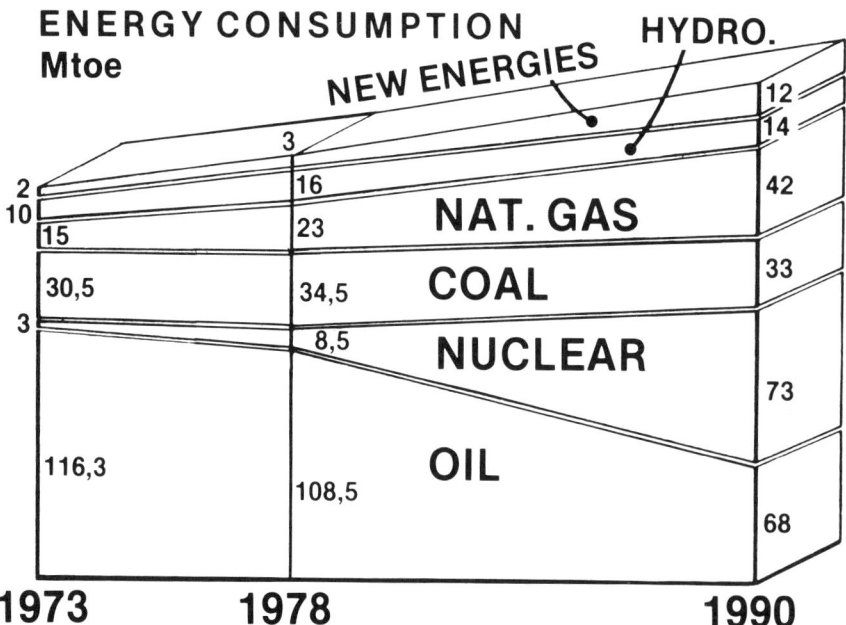

Figure 1-5. The Structure of Various Energy Sources by 1990

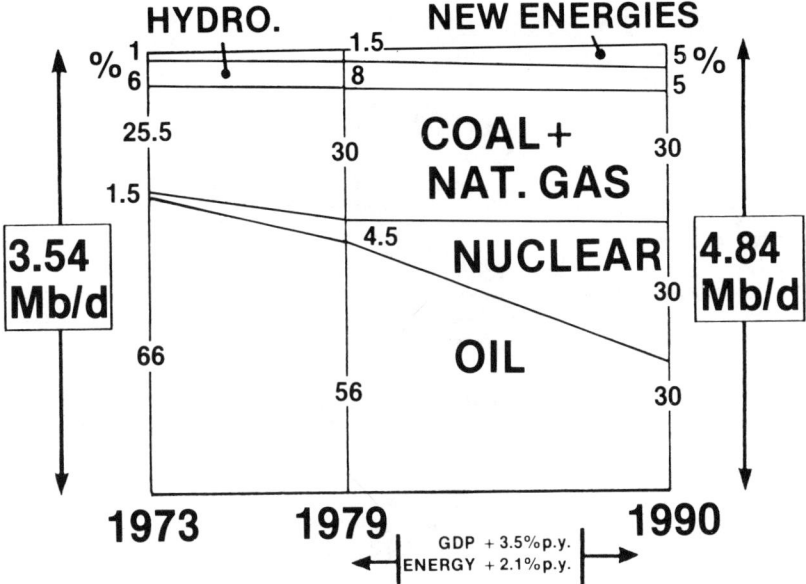

Figure 1-6. The Structure of Energy Consumption—Evolution, 1973-1990

pronounced changes between now and 1990. Figure 1-13 shows the decline in the importance of oil-based products, which currently account for more than 50 percent of total consumption in this sector, but which will be reduced to 10.5 percent in 1990. The figure also shows the emergence of new

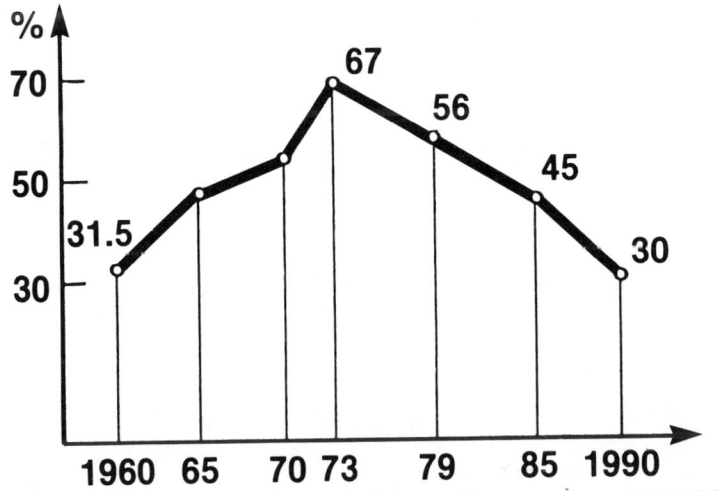

Figure 1-7. Oil Share in France's Energy Consumption, 1960-1990

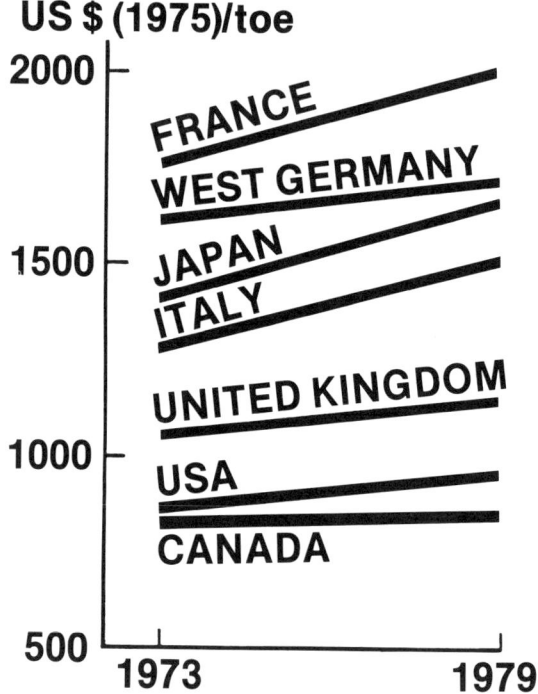

Figure 1-8. The Relation of GDP to Energy Consumption

forms of energy, which will represent 6 percent of the consumption of these sectors in 1990. The share of the market held by gas and coal will tend to increase slightly, and electricity will rise sharply to account for 58 percent of the sector's needs in 1990.

The implications of the leading aims of French energy policy for energy consumers will lead to changes in life-style, in the environment, in urban development, and in the basic design of buildings.

Transportation

Out of total oil savings of 60 Mt, the transportation sector will contribute 10 Mt. Figure 1-14 shows the trend in average fuel consumption by French cars, which already have a particularly low consumption level.

Energy Redeployment

Electricity: Nuclear Energy

Nuclear energy seems to be the most appropriate form of energy for producing electricity in a country like France, which is totally dependent on

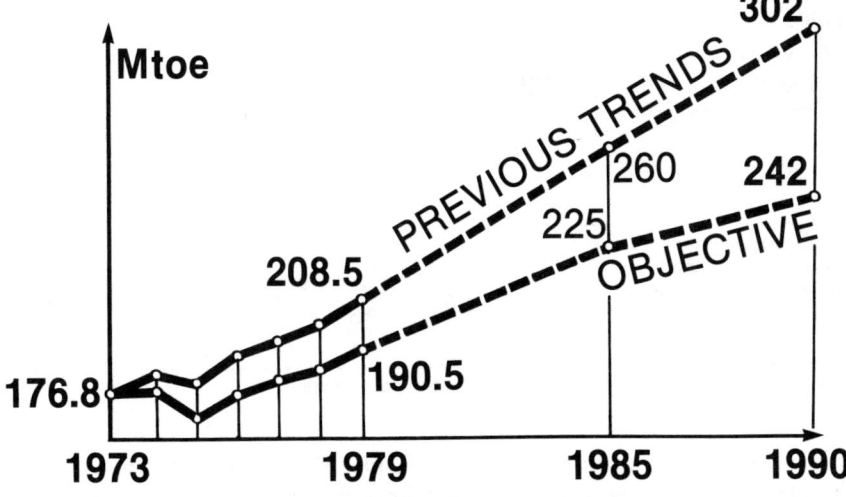

Figure 1-9. Energy Savings, 1973-1990

overseas countries for its oil, which has to import most of its coal, and whose hydroelectric equipment is at the saturation point. France has about 3 percent of world uranium reserves (100,000 tons). As part of an international undertaking, France has played a major role in building the Eurodif separation plant, with a capacity of 10.7 million separative work units

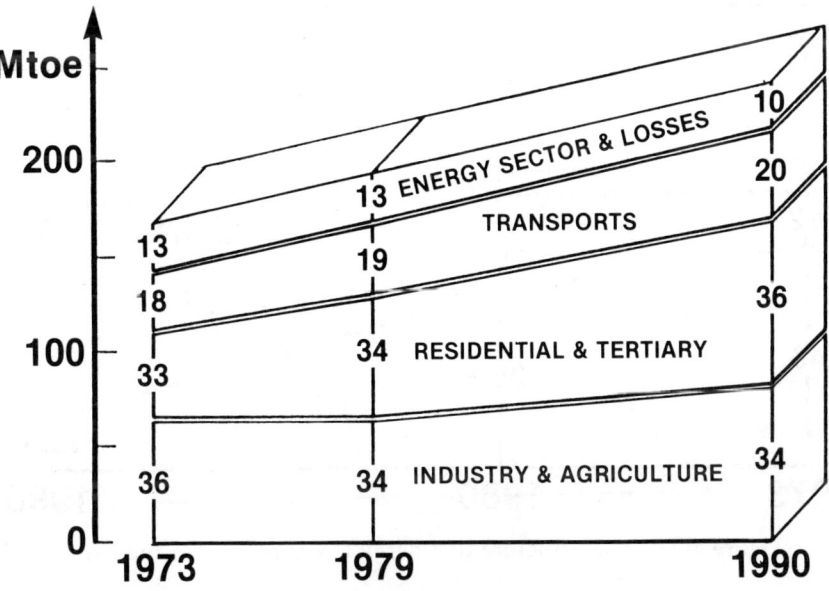

Figure 1-10. Breakdown of Energy Consumption

Table 1-3
Energy Savings
(Mtoe/previous trends—1973)

	1979		1990	
	In Mtoe	Percentage of the Sector[a]	In Mtoe	Percentage of the Sector[a]
Industry-agriculture energy sector and losses	5	5	24	18.5
Residential and tertiary	10	13	26	22.5
Transports	3	7.5	10	17.5
Total	18	8.5	60	20

[a]Percentage savings in the energy consumption of each sector.

(SWU), which will provide more than France's own needs for the coming decade, as well as a reprocessing plant whose capacity will increase from 500 tons in 1985 to 1,600 tons. Nuclear energy is also the most economically advantageous, as table 1-4 shows.

The production of electric power will rise by 5.6 percent per year from 1980 to 1990, compared with 6.5 percent between 1960 and 1980. The way that electricity will be produced in 1990 will be very different from the present method (figure 1-15). Nuclear energy will provide more than 70 percent of the electricity produced, 73 MtOE, compared with only 20 percent in

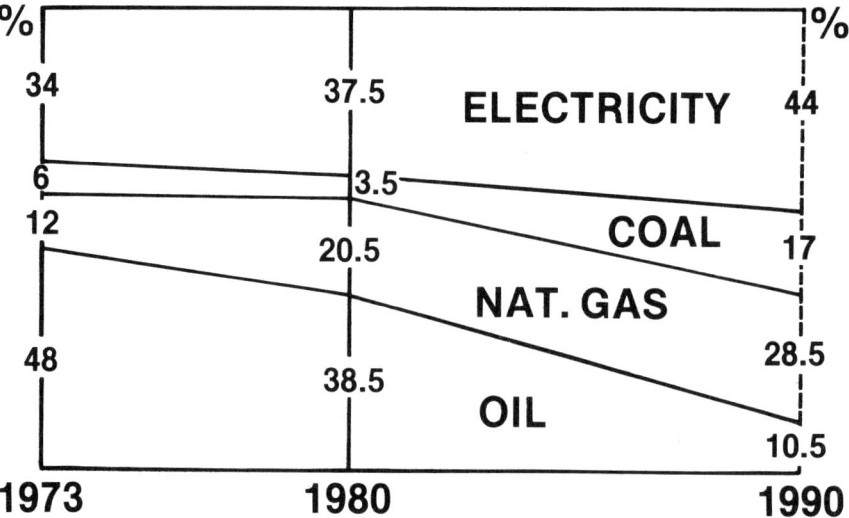

Figure 1-11. The Structure of Energy Consumption in Industry

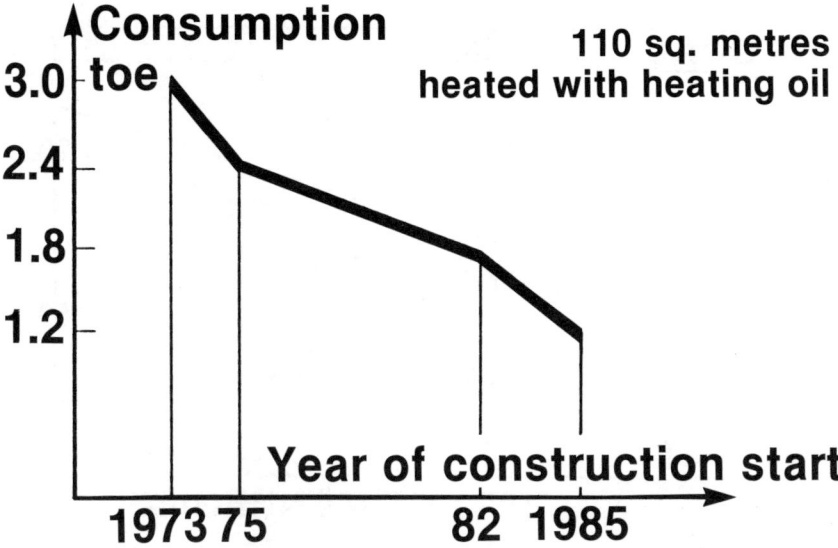

Figure 1-12. Energy Consumption in New Houses

1980. The first concrete results of the program set in motion in 1974 are now being felt. At the beginning of July 1980, nuclear plants were producing 25 percent of France's electricity needs.

As of 1981, 13,000 megawatts (MW) are operating, 32,000 MW are under construction, preparatory work is being undertaken for 22,000 MW, and a new plant goes on line every two months. Sixty-five months' lead time is required between the construction of the first concrete structure and the start-up of a 900-MW plant. This is a relatively short span of time, compared with that required in other countries. Does the French population accept such a program? The answer is yes. There have been only some local problems to date, although France may have been more fortunate than other countries in this respect. This is above all due to the following reasons:

> There has been very strong government commitment since the beginning. No hesitation, reluctance, or misgivings have ever been shown, even during the emotional period that followed Three Mile Island.

> There has been an extensive public-information campaign led by the government.

> There has been consensus among leading political parties.

> The French people, having suffered during two world wars, know firsthand the meaning of energy scarcity.

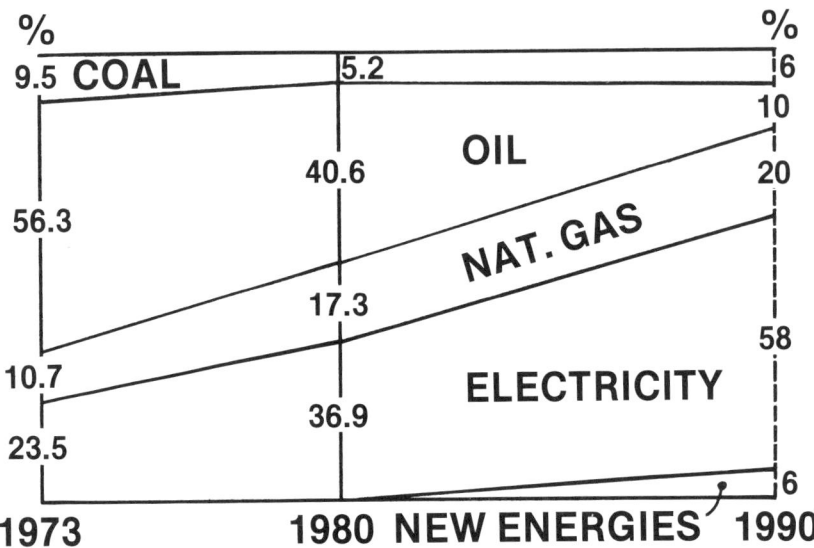

Figure 1-13. The Structure of Energy Consumption in Residential and Tertiary Sectors

Two other favorable factors are the existence of a single public utility, Electricite de France (EdF) for carrying out the program, and the backing of the French Atomic Energy Commission, which was created in the 1950s by Joliot-Curie and a staff of scientists who are famous and well known to the French.

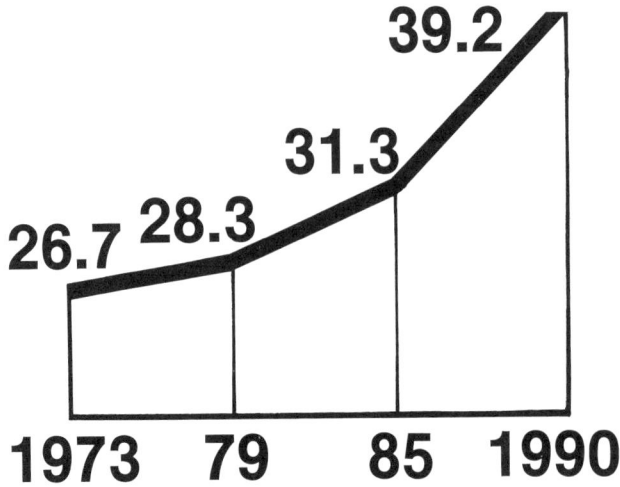

Figure 1-14. French Cars: Average Consumption, 1973-1990 (in miles per gallon)

Table 1-4
Comparative Generating Costs of Nuclear and Coal Power Plants
(1980—mills/kWh)

	Nuclear	Coal
Capital charges	16.1	12.3
Operating cost	6.1	6.8
Fuel cost	10.1	34.0
Total	32.2	53.1

Coal

The position of domestic coal as a proportion of French resources will inevitably decrease because of the small amounts of reserves (about 430 million tons, compared with 24 billion in West Germany and 45 billion in the United Kingdom) and the nature of the coal deposits (see table 1-5). The pattern of coal use will change substantially. Through coal consumption, industry will increase fivefold during this decade.

Gas

France will increasingly be forced to resort to imports of gas. These imports could represent more than 90 percent in 1990.

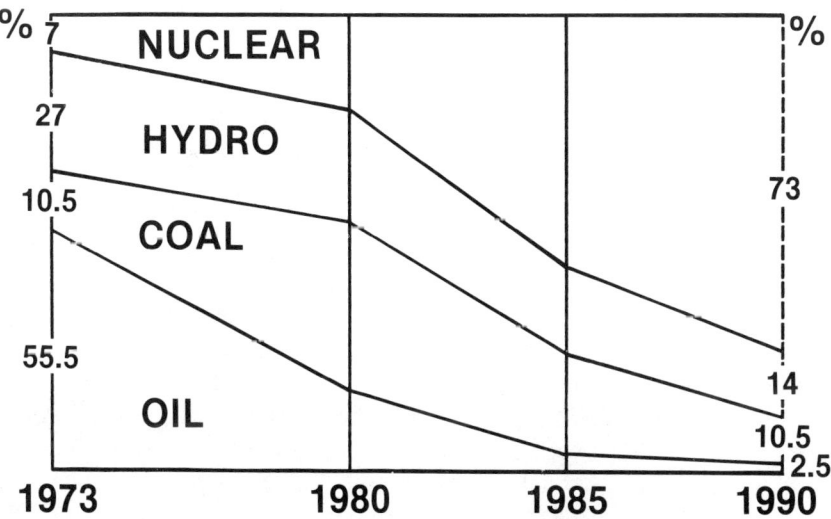

Figure 1-15. The Share of Nuclear in Electricity Generation

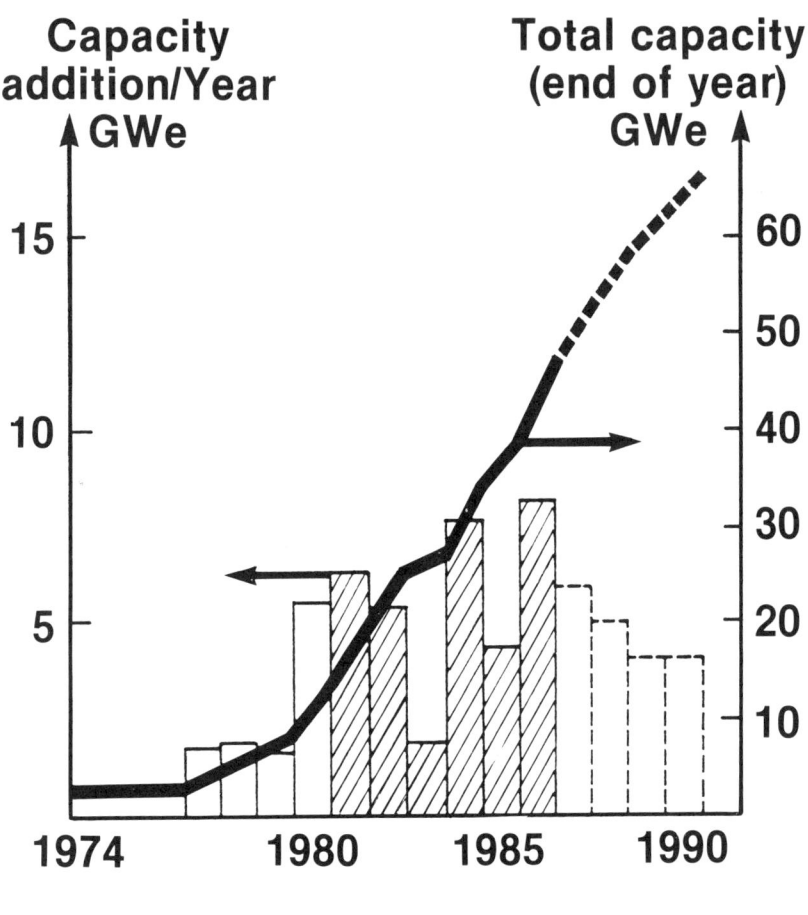

Figure 1-16. Growth of Nuclear Capacity

Renewable Energy Sources

By 1990 this contribution will have risen to around 24-26 MtOE as the result of a slight increase in hydroelectric power (14.5 MtOE) and a threefold increase in green energy (wood and agriculture), plus the new contributions of solar heating and geothermal energy (see table 1-6). Solar energy occupies a

Table 1-5
Changing the Pattern of Coal Use
(million tons of coal)

	1978	1990
Electricity generation	25.4	13
Steel works	13.1	15
Industry	2.7	16
Residential	5.3	5
Other uses	2.2	1
Total	48.7	50

leading position among the new forms of energy, both by the size of its overall contribution and by the diversity of its applications. Solar energy occupies a leading position among the new forms of energy, both by the size of its overall contribution and by the diversity of its applications. Solar energy can be exploited in various ways: in homes, by means of water heaters, solar heating, and solar architecture; and as biomass.

Oil: Taking Over from Oil

Oil conservation will remain significant since, in addition to the impact of energy conservation, there will also be the impact of the substitution policy (see figure 1-17). The essential feature of this figure is the steady decline of French oil consumption, which will drop from 100 Mt in 1985 to 68 Mt in 1990. Such a conclusion for 1990 is possible only as the combined result of energy savings amounting to 60 Mt, and of transfers to other forms of energy amounting to about 120 Mt. For France, therefore, all these factors together lead to a reduction not only in the share of oil in overall energy requirements but also in oil consumption in absolute terms—that is, oil for which the end use will be 70 percent devoted to transportation. Is this the ultimate outcome of this development? Perhaps not, if a country like

Table 1-6
Development of New Energy Sources
(Mtoe)

	1980	1990
Wood and biomass	3	7.5-9
Solar heating	—	1.3-1.5
Geothermal energy	—	0.8-1
Total	3	9.6-11.5

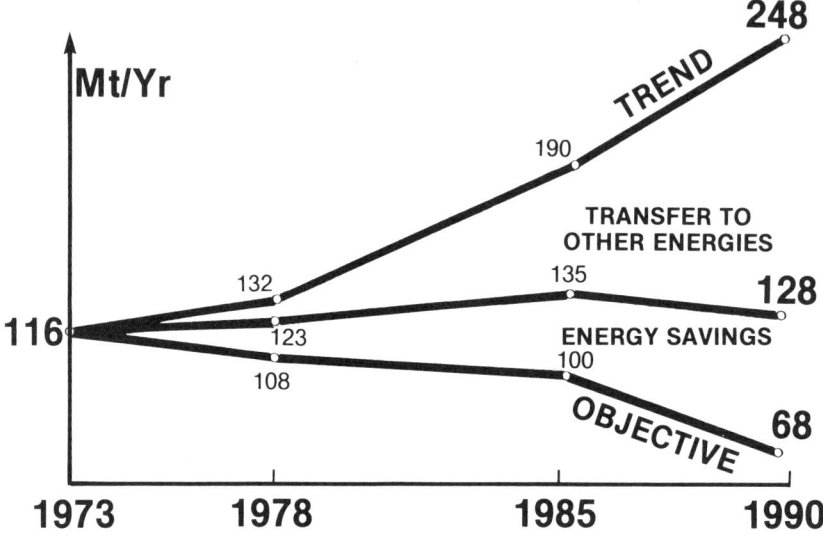

Figure 1-17. Oil-Consumption Objectives

France, well provided with agricultural resources, can find a portion of its alternative fuel sources in biomass.

The technical effort undertaken by France aims to gain mastery of the paths leading to an appreciable increase in available petroleum reserves (see table 1-7). This effort will concentrate mainly on offshore oil, which exists in amounts equivalent to onshore reserves; enhanced recovery, which could multiply the amount of available oil by 1.5 times; and heavy oil, which exists in amounts equal to those of conventional oil.

**Table 1-7
Mastery of Oil Technologies**

	Impact on Reserve Base
Offshore technologies	X 2
Offshore reserves $\simeq$ on shore reserves	
Enhanced recovery	X 1.5
Rate of recovery: Current $\simeq$ 30% Ultimate $\simeq$ 45%	
Heavy oil	X 2
Heavy-oil reserves $\simeq$ proved oil reserves	

Conclusion

French energy policy is thus a very deliberate one (see table 1-8), probably one of the most deliberate in the West, and perhaps even in the East. It is backed up by an overwhelming consensus in the country, for when we look at the French situation—3 percent of world consumption but 0.11 percent of world reserves—we have no choice. Yet such a strict policy must still be made to fit in with national growth and full employment.

Table 1-8
French Energy Investment Programs, 1980-1990

	Billion $/Year
Energy production	11.5-12.5
(from which nuclear)	(9-10)
Energy redeployment	7-8
Total	18.5-20.5

2 The Urgent Need for Realistic Planning Now

Mike McCormack

The election of a new administration by U.S. voters provides us all with an opportunity to reevaluate our positions and potential policies with the knowledge that the Reagan administration will emphasize energy production—production from all realistic sources—as well as conservation and that we can expect a constructive working relationship between the administration and the industries of the United States on which we must depend for energy production. At the same time, every realistic step will be taken to protect the environment and the health and safety of the people of the United States.

I consider this good news for the entire world because one of the major realities that almost all nations must face is the overwhelming need for a dramatic increase in energy production from realistic sources other than oil and, in the not-too-distant future, other than fossil fuels. This must be a cooperative venture among all nations, in which each will benefit from the success of all the others, regardless of differences in political, economic, or social philosophies.

One can easily make the case that for most of the world, including the Western industrialized nations and the less-developed and undeveloped nations, the most urgent and fundamental prerequisite for economic, political, societal, and military stability is the availability of an adequate supply of secure, reasonably priced energy. One can also make the case that the greatest assistance the Western industrialized nations can provide for the nations of the third and fourth worlds is to reduce their own imports of oil as much as possible during the 1980s and to provide the less-developed nations with realistic alternative energy technologies as these technologies become available in order to assist such nations in their own economic and industrial development while they minimize their dependence on imported oil.

All the people of the world are common victims of what we call the energy crisis. We are all in this together. Any single nation can improve its own situation by gaining energy independence for itself, but a sword of Damocles hangs over the heads of the world's people until we free ourselves from dependence on imported oil and then provide adequate energy from realistic sources within each nation to provide for growth and development and for increased standards of living for the people of the whole world. I am told that in Chinese the word *crisis* is made up of the combined symbols for *danger* and *opportunity*. Just as we are all collectively in danger, so may we all, if we work together to develop realistic energy programs, anticipate a wholesome and peaceful future for all the people of the world.

Many of us recall an old World War II slogan that went, "The difficult we will do immediately, the impossible will take a little longer." It is my conviction that we can start now to plan optimistically for the future and can assume a deep involvement by the United States in a collective, worldwide effort, first, to reduce our dependence on imported oil and, second, to provide energy abundance for the world.

Of course, we cannot accomplish all these goals during the 1980s. However, we can make a significant beginning. Indeed, we should be able to show progress within the next five years. A worldwide systems approach to integrated, realistic programs for energy production can help bring about a substantially more stable world in the year 2000 than exists today. Furthermore, I believe we can commit ourselves to the production of electricity and hydrogen from nuclear magnetic-fusion power plants before the year 2005 and can have substantial numbers of such plants on line by 2010. This will usher in a new era of unlimited supplies of clean, cheap energy for all humanity. However, the world must survive from now until the time when fusion plants start coming on line, and the next ten years will be the most dangerous.

Thus the challenge we face today is abundantly clear. We must overcome the paralysis that has gripped some of us during recent years and must develop energy policies and programs to eliminate waste, conserve energy, and protect the environment to the optimum extent and in every way practical, but also to produce from realistic sources the energy that we need—and will need.

If we do create such energy policies and implement the programs that will put them in effect, the world's people can have adequate energy and environmental protection, as well as economic stability. If we fail, the inevitable result will be worldwide economic, societal, and political catastrophe. But even setting out on such an undertaking will have a beneficial impact throughout the world.

First, there is an urgent need for realistic planning now—planning that treats the problems we face in their order of importance—as well as for a consideration of what is most practical to do first. It is obvious that the problem with highest priority is that of our dependence on imported oil and the economic, political, and military vulnerability that dependence creates today for the user nations, including the United States, on the one hand, and for the undeveloped nations on the other.

In this case, conservation plays its most important role. The people of the United States have accomplished a great deal in reducing consumption of petroleum products, especially of motor-vehicle fuels and heating oil. This has been accomplished through a heavy emphasis on conservation, involving, for instance, the conversion of the U.S. automobile fleet from large, inefficient vehicles to smaller vehicles with about twice the gasoline mileage efficiency of the average 1973 automobile. This conversion is, of course, not yet complete; but the progress we have made, along with the tendency of people in the United States to drive fewer miles because of both the desire to conserve and the cost of fuel, has made it possible for the United States to reduce imports of petroleum and petroleum products substantially. In addition, the United States has made significant progress in reducing the consumption of energy in industrial operations and in residences. The campaign to insulate homes and to avoid heating them to extremes in winter and cooling them to extremes in summer has resulted in a substantial reduction in energy consumption in U.S. residential use. I hope that every advanced nation will undertake similar programs of energy conservation and that throughout the world new industrial development and new residential construction will take into consideration the need to conserve energy.

This is, of course, a new phenomenon in the United States. Both the United States and Canada have developed vast areas of land during the last 150 years under conditions in which energy was generally cheaper than building materials, and transportation was sufficiently inexpensive to allow cities to sprawl to the point at which transportation became an everyday necessity. We are in the process of a moderate change in our style of living; but we will not reduce our per capita consumption of energy to that of the smaller countries of Western Europe unless we consolidate our population within much smaller land masses, as is the case in Western Europe, and crowd our people into smaller living areas in our most energy-efficient homes. Of course, no such unrealistic schemes are contemplated; and it is easy to project a point at which further reductions in per capita energy consumption will have unacceptable societal consequences.

Hence conservation alone will not solve U.S. energy problems. Neither will the development of new energy sources, such as those involving solar energy, geothermal energy, the conversion of wastes such as municipal garbage and other wastes to energy or fuels, or the production of fuel such as alcohol from agricultural crops. The most optimistic and honest evaluation of the potential of all these new energy sources combined, along with new hydroelectric generation within the United States, indicates that at most they will produce no more than 10 percent of the energy the country will consume in the year 2000. Furthermore, even if we are successful in meeting former President Carter's projection that we reduce, through conservation, our energy consumption by one-third by the year 2000, the United States must still essentially double its present energy production capacity during that period just to reach that goal. A similar challenge faces most of the nations of the industrialized world, and this situation is exacerbated by the near certainty of a reduction of the availability of petroleum and natural gas throughout the world in the coming years.

Even the availability of petroleum for purchase from the Organization of Petroleum Exporting Countries (OPEC) does not provide us with the necessary security. On several recent occasions Dr. John Sawhill, deputy secretary for the Department of Energy, projected that the Organization for Economic Cooperation and Development (OECD) nations (the United States, Canada, Japan, Australia, New Zealand, and Western Europe) will be importing about 28 million barrels of petroleum and petroleum products a day in 1985. At $35 a barrel, this would mean that these OECD nations would be exporting about $350 billion a year to the OPEC countries. It is obvious that this continuing drain of cash from the West to the OPEC nations would constitute an unacceptable economic situation.

Today, the $90 billion per year the United States is paying for imported oil, along with increased costs for petroleum and petroleum products within the United States, constitutes by far the most significant inflation factor in the U.S. economy. A study by the Congressional Budget Office indicates that of recent increases in the U.S. cost of living, 25-30 percent is directly, and perhaps an equal percentage indirectly, attributable to the increased cost of imported oil.

Another problem associated with the energy crisis that is not generally appreciated in the United States has to do with many of the undeveloped and underdeveloped nations of the world. The collective debt of these nations for the oil they have imported in recent years is so great that they have no way to repay it. They resent the high rate of consumption of OPEC oil by Western nations, especially by the United States. This resentment is sowing the seeds of more serious problems between third-world nations and the United States in the not-too-distant future. The hardship of the Third World is real and visible. For instance, I am told that agricultural machinery in some countries has been left idle because there is no fuel and no anticipation of obtaining fuel in the near future to operate the equipment.

I have not focused on the future economic and political problems that will flow from a shortfall of available energy supplies in any country. It may suffice to observe that there is now, and always has been, a close interdependence between energy consumption, gross national product, employment levels, and standard of living. This is especially true in the Western industrialized nations.

At a recent hearing before the House Subcommittee on Energy Research and Production, which I chair, Dr. Ronald Ridker of Resources for the Future provided us with the following carefully formulated projections concerning the worldwide energy situation:

1. Assuming that economic growth continues, commercial energy requirements are likely to increase dramatically in the next forty to fifty years, doubling by the year 2000 and doubling again by 2025. Per capita

consumption will also rise in all regions studied, in some places quite rapidly. If supply cannot keep up, economic growth could be substantially less than assumed in these projections.
2. Energy use per dollar of GNP is likely to fall in the United States and other already industrialized and urbanized countries, but is expected to rise during the next twenty years in countries less advanced in these dimensions. Thereafter, these countries also show a decline. The net effect for the world as a whole is a slow but continuous drop in energy use per dollar of GNP. In other words, the whole world will become more economical in its use of energy, but only very slowly because of the increased demands in countries that are rapidly industrializing.
3. Growth in income per capita is even more important than growth in population alone in explaining the increased demand for energy. Other factors, such as urbanization rates and prices, are also more important, at least during periods when they are changing significantly. Global population growth is of profound, ultimate importance; but it occurs too slowly and its path cannot be altered rapidly enough to make it as important as these other factors within the next two or three decades.
4. Given the size of the world's petroleum resources, the demand for liquid fuels in coming decades will have to be met by the development of synthetics and by substitution of nonliquid fuels—for example, production of electricity from other fossil fuels and nuclear power. If these technologies do not come along at a reasonable cost to fill the gap, and if current thinking that global petroleum production is already near its peak proves to be correct, the world's nations could be on a dangerous collision course.

Clearly, there is an urgent need for realistic energy planning on an international scale, beginning as soon as possible. This year's summit conference in Vienna emphasized the need to reduce Western dependence on imported oil through a much heavier reliance on coal and nuclear fission, with a special emphasis on the production of synthetic fuels from coal; and with environmental protection and appropriate consideration of public health and safety in producing energy from these two major sources. I believe that the conclusions of the Vienna conference must be implemented aggressively and immediately. To the extent that it is practical, oil-burning electric-generating plants, especially for baseload electricity, should be converted from the use of petroleum and natural gas to the use of coal. This should be true throughout the world and should apply regardless of the source of coal. I do not make this suggestion for the sake of increasing U.S. exports. The origin of the coal—the exporting country—is much less important than the need to relieve dependence on OPEC oil.

However, we should not assume that the United States can mine and export huge quantities of coal, perhaps doubling its coal production during the next fifteen years, without a great deal of planning, the hiring of several hundred thousand new coal miners, the rebuilding of the U.S. transportation system to haul coal to ports, and the investment of a huge amount of capital to fund such major construction projects. In addition, it is essential that as much coal as possible be converted to clean gaseous or liquid fuels before consumption. For export, conversion of coal to liquid fuels is highly desirable but would involve the addition of a new major industrial infrastructure for that purpose, itself costing hundreds of billions of dollars.

Some planning must be unique to each country. For instance, the United States must seriously consider a rational reorganization of its federal agencies to optimize the development, management, and regulation of its energy resources and technologies. Such a reorganization would involve the Department of Energy, the Department of Interior, the Nuclear Regulatory Commission, the Environmental Protection Agency, and the Council on Environmental Quality, among others. Today, these agencies often work at cross purposes. This must be corrected, and an efficient system that emphasizes energy production, along with human health and safety and environmental protection, must be established. Several regulatory laws must also be reformed to make them compatible with a realistic energy-production policy. These laws include the National Environmental Policy Act, the Surface Mining and Reclamation Act, the Clean Air Act, the Clean Water Act, the Federal Water Pollution Control Act, the Toxic Substances Control Act, the Nuclear Power Plant Licensing laws, and the Nuclear Export Control Act. Although these laws have the commendable purpose of protecting public health and safety and the environment, they contain provisions that will continue needlessly and seriously to delay energy and other industrial development in the United States. I propose that all these laws be carefully rewritten so that they provide the protection that was intended when each was originally enacted, but so that they cannot be used as weapons of harassment and obstruction by those who would, without constructive purpose, interminably delay or prevent the construction of essential energy and other industrial facilities in this country.

One prerequisite to implementing nuclear energy-production programs is an extensive campaign of public education, providing factual and understandable information about nuclear energy so that people can accept it for what it is: the safest, cleanest, cheapest, most reliable, and most environmentally acceptable significant energy source available during the balance of this century and, indeed, until we start building commercial nuclear fusion plants in the twenty-first century.

We must, of course, institute programs to minimize the potential for any diversion of nuclear material for any illicit purpose. Although I believe the actual threat of a diversion from the nuclear energy cycle to the fabrication of weapons has been vastly exaggerated, I do think it is possible to take certain institutional and technical steps to prevent such diversion.

The Urgent Need for Realistic Planning Now 25

The United States must take the lead in establishing a multinational nuclear-fuel-reprocessing center. This can best be accomplished by buying the Barnwell facility in South Carolina, converting it to coprocessing, and adding to it an advanced fuel-fabrication facility and a waste-glassification facility. Such a center would be managed by the nations it serves and would produce only new fuel elements and canisters of glassified waste. New fuel bundles would be slightly irradiated before being shipped from the facility, thus requiring heavy shipping casks for moving fuel both to and from power plants and requiring heavy remotized equipment and shielded facilities for handling even new fuel elements. Such a recycling system and such procedures would reduce almost to zero the potential for the diversion of any fuel material to any unauthorized purpose. It would also encourage all but the largest and most advanced nations to lease fuel elements from the multinational agency (under International Atomic Energy Agency monitoring) rather than trying to develop independently the expensive and difficult technologies associated with fuel reprocessing and recycling.

Multinational fuel-recycle centers would provide a realistic and far more acceptable deterrent to nuclear-weapons proliferation than is provided by the present unilateral United States Nuclear Export Control Act, which is both resented and ineffective. Each international center should provide a fuel bank at some other location that would contain an assured supply of fuel for one year for each participating nuclear power plant served by that center.

There are many areas for further international cooperation. We must accelerate research and development associated with increasing the life of conventional nuclear power plants and increasing the fuel utilization—the time in reactor—of nuclear fuel. We must develop technologies for reducing radiation levels associated with midlife maintenance of nuclear plants; and we must, of course, take whatever steps are indicated to improve even more the safety of existing and future plants. We must also move forward with the breeder program, which is essential.

In addition, the time has come to move into the engineering phase of nuclear magnetic fusion. This can now be done with confidence. The new fusion-engineering law is the most important energy initiative that any country could undertake. There is great potential for international collaboration in this effort as we move toward a final solution to our energy shortages.

It is important to believe that we *can* succeed. With a new administration in the United States, the nations of the world can move forward together, combining their efforts in rational programs for energy production involving realistic sources, such as coal and nuclear fission now and nuclear fusion when it becomes available. This venture should be characterized by the optimism and purposefulness we have shared at other times in pursuing common goals. It is important to realize that there are no significant technical problems facing us that do not have solutions. We must provide the leadership for political solutions, taking advantage of existing technology so that the world's people can have adequate energy production, economic stability, and international security, as well as protection of the environment and human health and safety. This goal is now within reach.

3   Getting from Here to There: The Energy Road Ahead

M.E.J. O'Loughlin

This chapter deals with likely energy supply and demand in the years ahead. The future, for a number of reasons, now seems both more hazardous and more exciting than ever. In today's world, untoward events can happen at a moment's notice, as the conflict between Iran and Iraq demonstrates. Such events often follow a trajectory of their own, upstaging the most carefully developed forecasts.

Short-term projections are particularly vulnerable. Today's glut becomes tomorrow's scarcity; almost as quickly, optimism—perhaps unwarranted—is transformed into pessimism—perhaps equally unwarranted. What is needed is the best possible understanding of underlying trends. We need some sense of what the energy landscape will look like ten and twenty years hence, so that we can start planning for tomorrow, today. A great deal of time is required—ten to fifteen years, in fact—to develop large-scale new energy sources. The underlying trends indicate that we will need a variety of new sources—coal, nuclear, synthetics, and eventually such renewable sources as solar and fusion—each in its proper time and role. It is important to have a sense of how all these pieces can fit together, for they would be required even if the consuming nations were not in danger of crude-supply disruptions, even if 50 percent of the world's petroleum reserves were not situated in the Persian Gulf where some of the nations are unstable.

Before proposing any solutions, we must analyze the problem. The question is, How do we get from an era of extensive dependence on conventionally produced oil and gas to that time in the next century when, presumably, the world will rely predominantly on renewable and essentially nonexhaustible forms of energy. What is the path that gets us from here to there?

As figure 3-1 shows, our view of the future keeps getting revised. For example, between 1973 and 1975 events put quite a dent in our earlier projections of world energy demand. And I cannot say with assurance that what we project at this time will stand the test of time any better. I would like to emphasize that the outlook presented in this chapter is not intended as a prediction. Rather, it is one set of plausible projections about supply and demand to the year 2000; and as with all such exercises, it assumes away certain awkward developments that could stand in its way—such as unusually deep economic depression and major political upheavals.

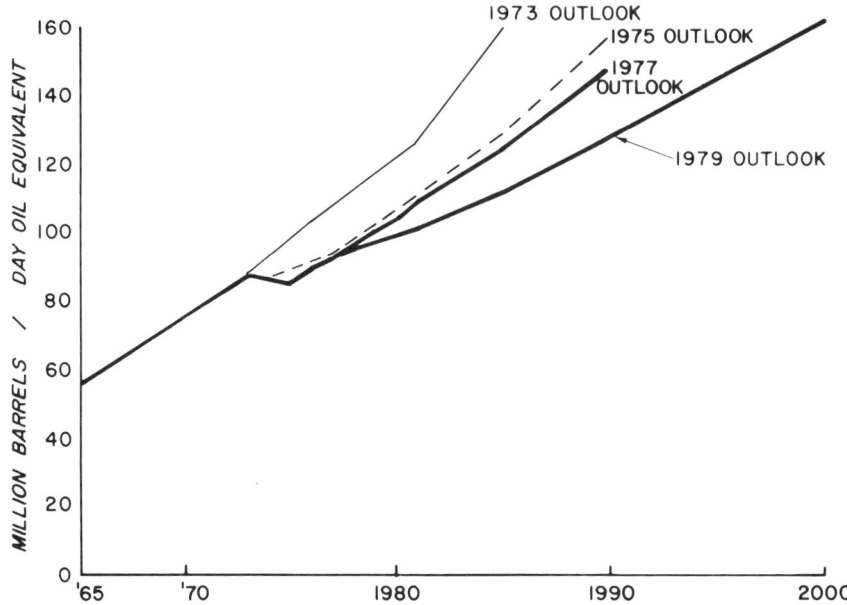

Note: Excludes Communist areas.

Figure 3-1. World Energy Demand: Changing Perceptions

Let me also note that we will soon release an updated outlook. But the numbers and the trends are not significantly different from the ones to be given here. For example, there will be further slight markdowns in world energy demand and in world economic growth. Directionally, however, things will be much the same.

As figure 3-2 indicates, total demand for energy in the noncommunist world is expected to grow over the next several decades less than half as fast as it did before the Arab oil embargo. During that earlier period, the growth rate was about 5.5 percent a year, compared with 2.5 percent during the foreseeable future. Even so, by the year 2000 world energy demand is projected to be some two-thirds greater than today. We expect the U.S. share of total energy usage to decline from its current level of 41 percent to less than one-third by the turn of the century. The European industrial countries' share will dip slightly, whereas Japan will remain unchanged. In contrast, those grouped as "others"—mainly the developing countries—are projected to increase their demand from 19 percent to one-third of total energy by the year 2000. Breaking down these others still further, the

The Energy Road Ahead

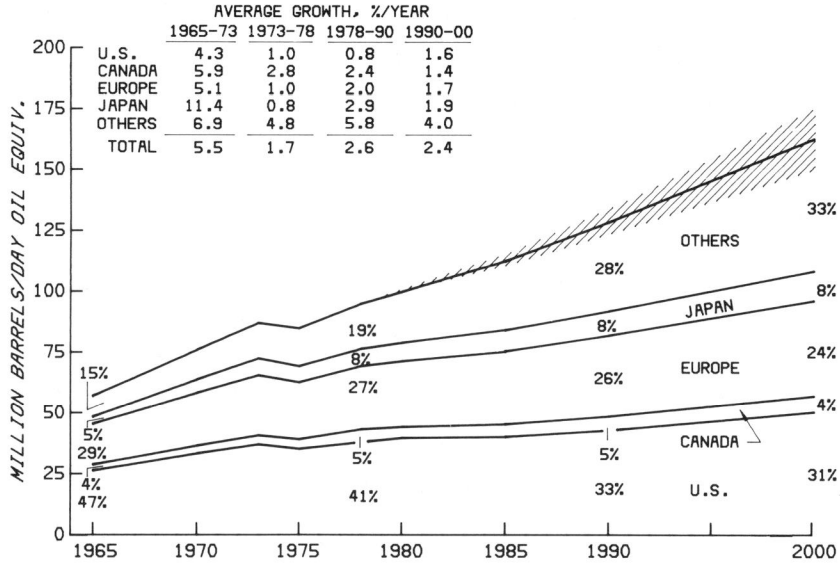

Note: Excludes Communist areas.

Figure 3-2. World Energy Demand

energy-exporting countries will probably increase their usage from 8 to 16 percent by the century's end, whereas the energy-importing less-developed countries (LDCs) will increase their share from 11 to 17 percent.

Consider the kinds of energy the world is likely to be using in the next few decades. As you can see from figure 3-3, by the beginning of the twenty-first century, conventional oil and gas will supply much less of total world energy than they do today. Oil will decline from 54 percent to about 37 percent by the year 2000, and natural gas will go from 18 percent to 16 percent, despite an expected 50-percent increase in natural-gas production. Part of the burden once borne by gas and oil will be shifted to coal and nuclear. Coal is projected to break out of its recent pattern of negligible growth as it jumps from supplying 18 percent of total energy currently to 24 percent two decades hence. Nuclear's share will rise to 10 percent, compared with only 3 percent today. Almost three-quarters of nuclear growth in the 1990s will be outside the United States, and the number of plants needed to come on line during that decade is not much higher than the rate for the 1980s.

Synthetics could be another significant contributor to world energy needs in the next twenty years. They could supply 4-5 percent of total demand by 2000. In fact, they are expected to achieve the most rapid annual growth of all fuels during the balance of this century.

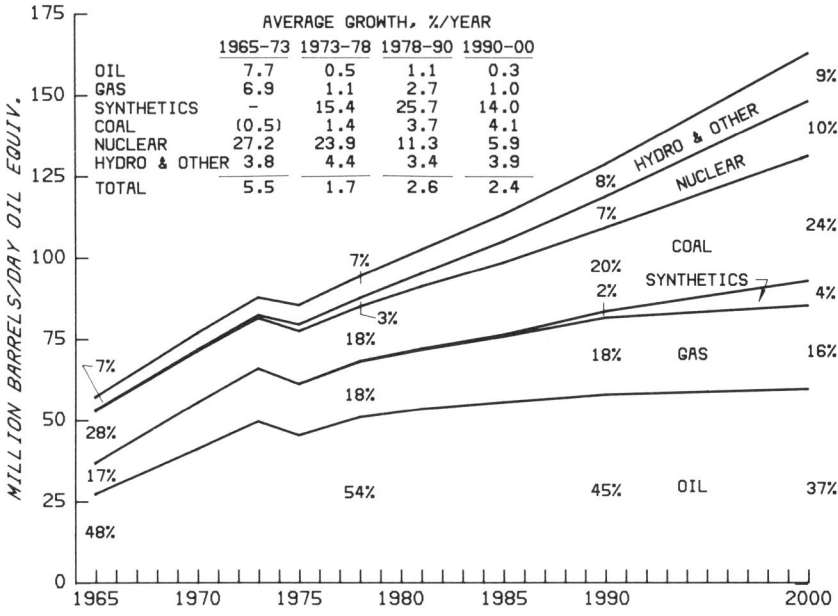

Figure 3-3. World Energy Supply

Hydropower's overall share is projected to increase only modestly in the next twenty years, and solar power will probably supply only about 1 percent of world energy by the year 2000. Although solar may be important in the future, even to achieve the fairly modest impact we project, about 15 million homes in the United States, Europe, Canada, and Japan would have to have solar domestic hot-water systems, and about 7.5 million homes would need solar heating systems. This is a tall order, considering where we stand today.

Turning now to coal, we see from figure 3-4, as already noted, that world demand is expected to grow steadily until the end of the century, in contrast to the nearly flat demand curve since 1965. In terms of oil equivalencies, coal will increase from some 17 million barrels a day currently to nearly 42 million barrels a day by the year 2000.

In particular, coal use in the industrial sector is expected to expand because of higher oil prices as well as U.S. Legal requirements. Beginning in the late 1980s, demand for coal to be converted into gas or liquid synthetics will probably accelerate. In two decades this form of usage is expected to make up about 12-15 percent of consumption, most of it in the United States.

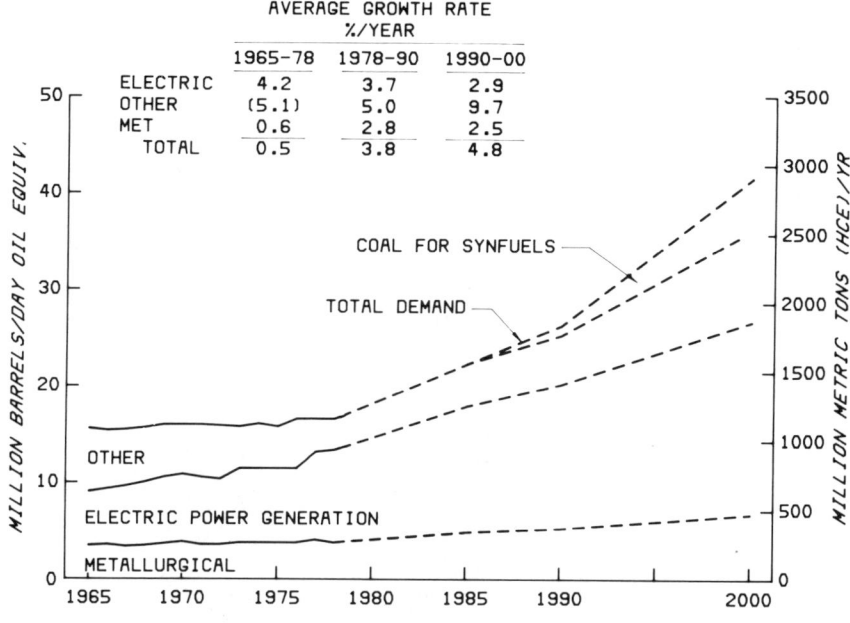

Figure 3-4. World Coal Demand

Shifting to the supply side of the equation, the United States continues as the world's major coal producer. Half of world output is expected to come from this country by the turn of the century—the equivalent of nearly 21 million barrels of oil a day out of some 42 million barrels. Although U.S. reserves are large enough, U.S. environmental regulations could add materially to the cost of developing and operating new mines. Also, Transportation facilities will have to be expanded both to move coal domestically and to ship it abroad, especially to Europe and Japan. In Latin America, Africa, and the Far East, we expect to see expanded production capacity aimed at export markets. This would mean major increases in the amount of coal entering international trade.

The most striking feature of figure 3-6 is the extent to which the less-industrialized developing countries are expected to increase their demand for oil over the next two decades. In fact, it will more than double, so that by the end of the century these nations will account for 43 percent of world consumption. In contrast, demand in the industrialized countries is not expected to grow during the outlook period. It will even show a slight reduction by the year 2000, when the developed nations will account for only 57 percent of total needs. The U.S. share will drop from 37 to 25 percent of

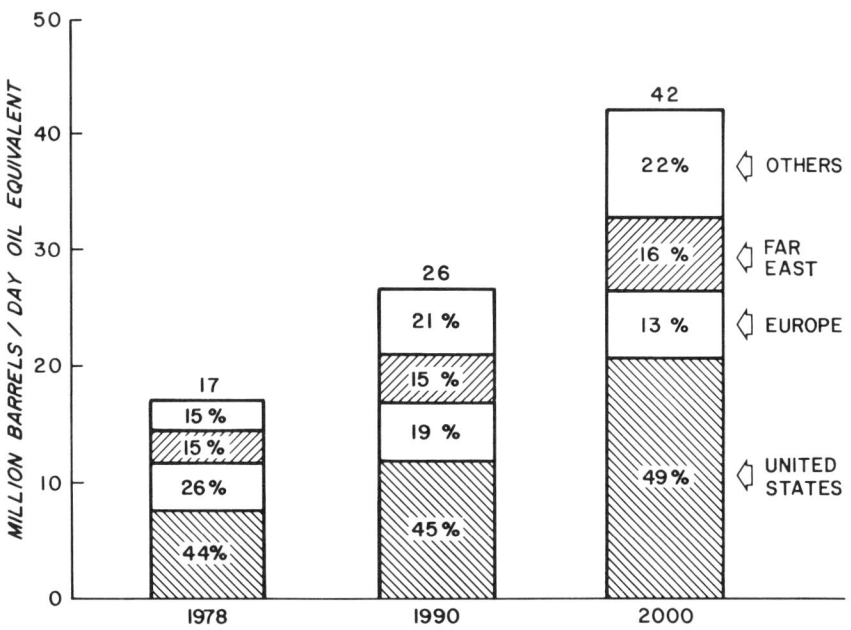

Note: Including net imports from Communist areas.

Figure 3-5. World Coal Supply

world demand. As the figure shows, today three-quarters of the world's total oil consumption is in the United States, Japan, and Europe.

The table accompanying figure 3-6 shows how much of their oil demand the United States and the European countries will be able to satisfy from their own conventional reserves. Although they provided the United States with nearly 80 percent of its oil needs in 1970, by the year 2000 they will yield a significantly lower percentage. The balance will either be imported or come from synthetics manufacture. In Europe, domestic production—largely from the North Sea—will provide a much greater share of local needs. It will rise to 28 percent of demand by 1990, compared with 3 percent ten years ago.

Figure 3-7 deals with the countries that will be providing future oil supplies. Oil here includes crude, liquids recovered from natural gas, and liquid synthetics. OPEC production is assumed to remain almost constant over the outlook period. Obviously, however, predicting OPEC production levels is hazardous. Some exporting countries want to limit production because they

The Energy Road Ahead

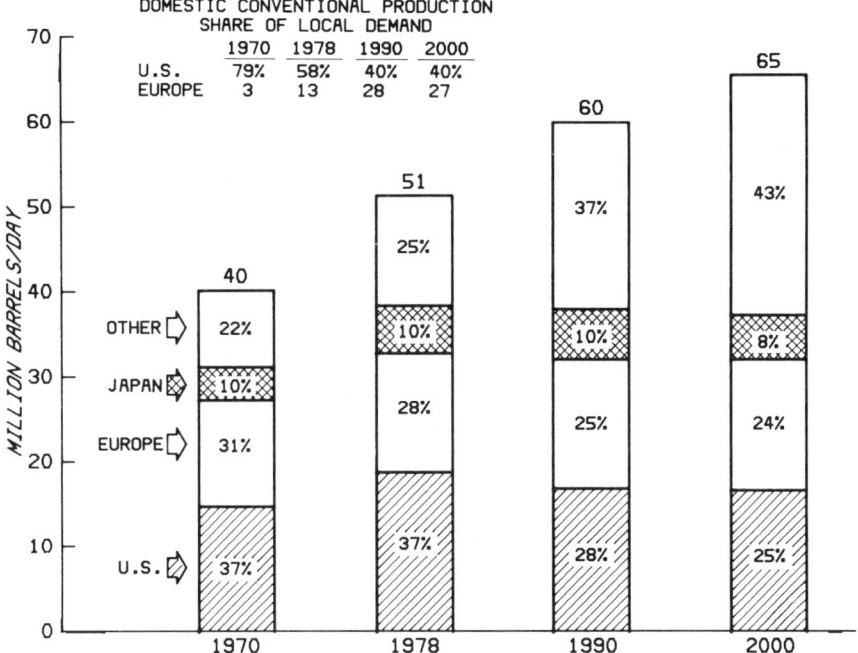

Note: Excludes Communist areas.

Figure 3-6. World Oil Demand

expect the rate of discovery to be less than withdrawals, meaning that reserves will continue to decline as the value of the oil in the ground increases.

Elsewhere, in the United States and Canada, conventional oil production is expected to decline; however, it will increase in Europe as North Sea reserves are developed. The biggest foreseeable increase will be in other non-OPEC countries, where production is projected to triple over the next two decades. Mexican production especially is likely to play a big part here. As for net exports from the communist countries, dwindling volumes from the USSR are projected to be gradually offset by greater amounts from the People's Republic of China. By the year 2000 liquid output from synthetics plants could rise to the equivalent of around 6-8 million barrels of oil a day.

But as the bar on the right of figure 3-7 shows, the synthetics volumes are virtually all from plants yet to be built. Furthermore, roughly 35 percent of conventional oil is expected to come from reserves that have not yet been found. Such is the highly tentative nature of the world's future oil supply.

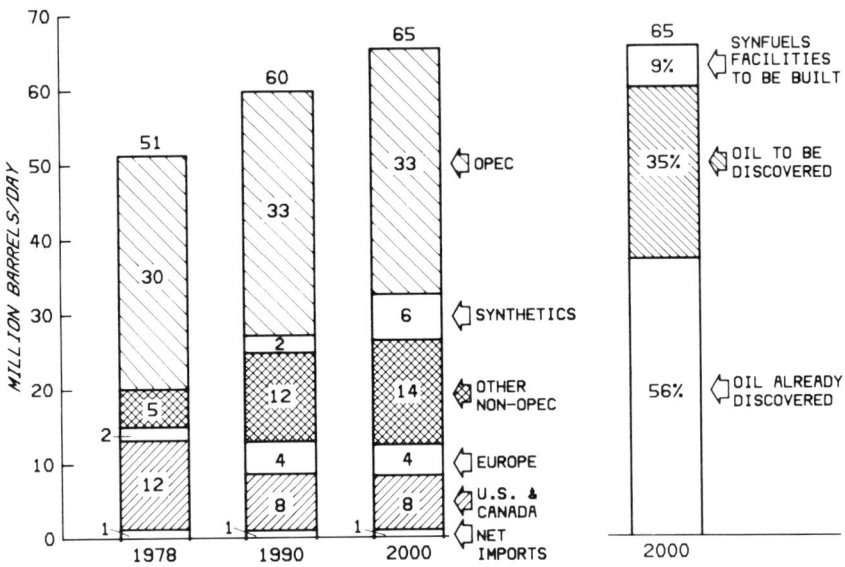

Note: Excludes Communist areas. Totals may not add exactly due to rounding.

Figure 3-7. World Oil Supply

It is difficult to generalize about natural-gas use in the major consuming countries because they have followed very different patterns, reflecting the pace at which local supplies have been discovered and developed. Substantial international shipments of this fuel have begun only recently. Now, however, a number of large-scale projects are underway that would enable natural gas to be moved from one part of the world to another.

As figure 3-8 shows, growth in gas consumption will largely depend on what happens in the developing countries, where a number of projects for using local gas production are in the works, especially in the Middle East and Latin America. U.S. gas production peaked in 1973 and has been declining ever since. In Europe consumption has grown rapidly since the early 1960s, when the Dutch reserves were discovered. In Japan gas use is based almost entirely on imports, and these are expected to continue increasing.

Figure 3-9 illustrates the United States' future energy outlook. However, note that perhaps the most significant element is not incorporated in the figure—namely, the effect of conservation. By the turn of the century the energy used to produce each unit of GNP should be 30 percent below the levels that would have resulted if pre-1970 patterns had continued.

In line with the world situation, coal and, to a lesser extent, nuclear will be the growth fuels for the next twenty years in the United States. Nuclear's

The Energy Road Ahead

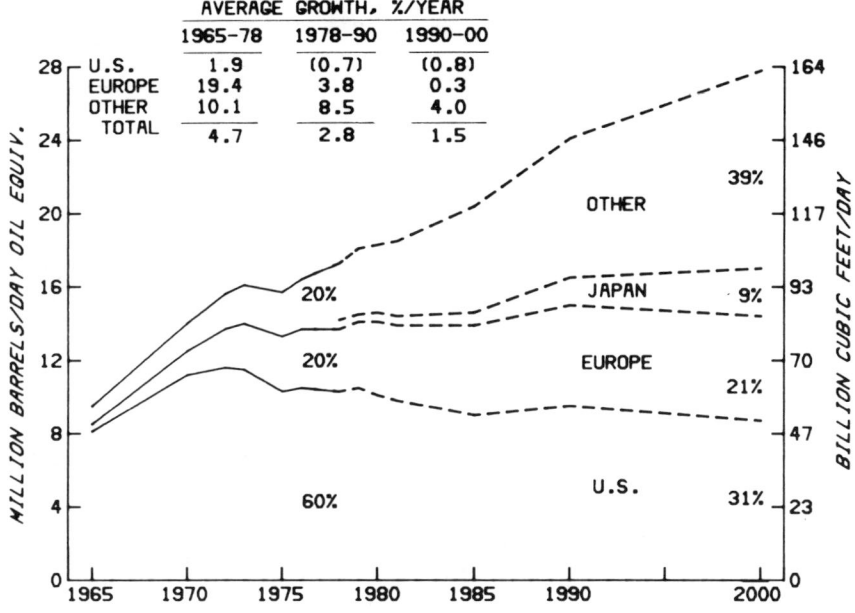

Note: Excludes Communist areas.

Figure 3-8. World Gas Outlook

indicated growth rate of 9 percent a year in the 1980s includes reactors that are either on order or under construction. But to achieve a 4.5-percent annual growth rate in the 1990s, new orders will be needed. Hydropower, geothermal, alcohol from biomass, and solar energy are considered together and are expected to grow somewhat.

Natural-gas production is projected to decline, and domestic oil production will gradually go down from its current level of 10.3 million barrels a day to between 6 and 6.5 million barrels in the 1990s. Synthetics, from both coal and oil shale, will have a direct impact on the level of U.S. energy imports. A domestic synthetics industry could produce the equivalent of 4-6 million barrels of oil a day by the year 2000, and perhaps 15 million barrels per day ten years after that. Figure 3-9 shows that by the end of the century, oil and gas imports could represent more than half the United States' need for these fuels. The hatched area represents the potential for reducing imports if synfuels development is accelerated. Because of lower demand and slightly higher petroleum supplies, however, our new outlook will probably show a somewhat smaller import requirement, although it will still be a sizable amount.

Clearly, the world is in a transition away from the age of petroleum and toward an age of energy diversity. But this changeover will not take place overnight. Oil will still be the dominant fuel by the turn of the century, even

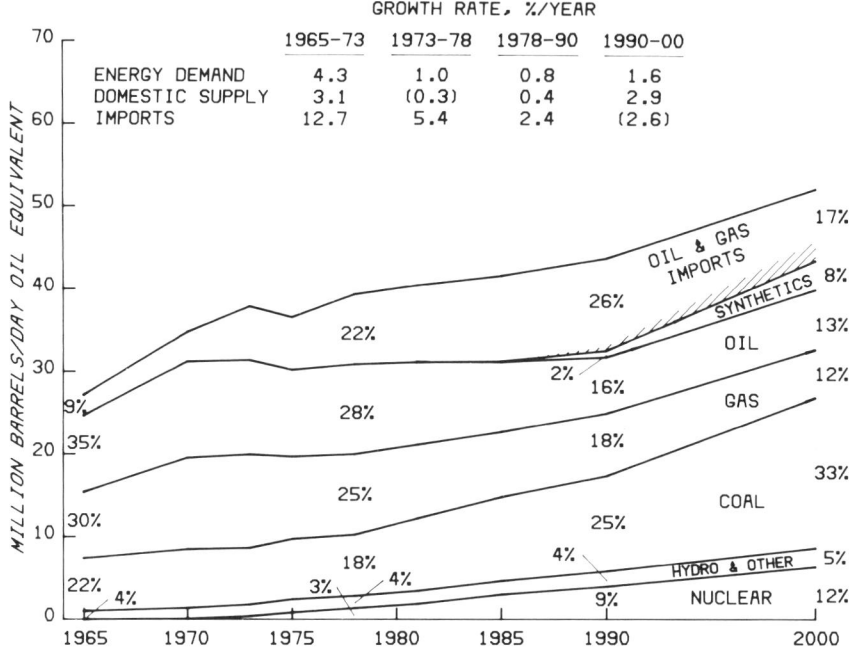

Figure 3-9. U.S. Energy Outlook

if, as mentioned earlier, it provides a smaller percentage of total world energy needs. As one of the figures indicated, by the year 2000 the world will probably be consuming greater volumes of oil than it does today—14 million more barrels a day, in fact.

Figure 3-10 shows that not only oil and gas sometimes take a lot of time and money to develop—so do the alternatives. For example, some seven to ten years would be needed to go from drawing board to functioning reality for a single synthetics plant producing 50,000 barrels a day. What is more, that plant would cost about $3-$4 billion in 1981 dollars. If we multiply that by what it would take to create a major synthetics industry in this country, we rapidly reach the financial stratosphere. Conservation in its various forms is of course likely to show results sooner, but its dimensions are uncertain.

In all parts of the globe, both the public and the policymakers appear to be increasingly aware and informed regarding the issues. Perhaps the best evidence of changing attitudes is all the research and activity on alternative sources. For example:

> In Brazil, alcohol made from sugar cane is expected to supply a hefty percentage of that country's motor-gasoline requirements by 1990.

The Energy Road Ahead

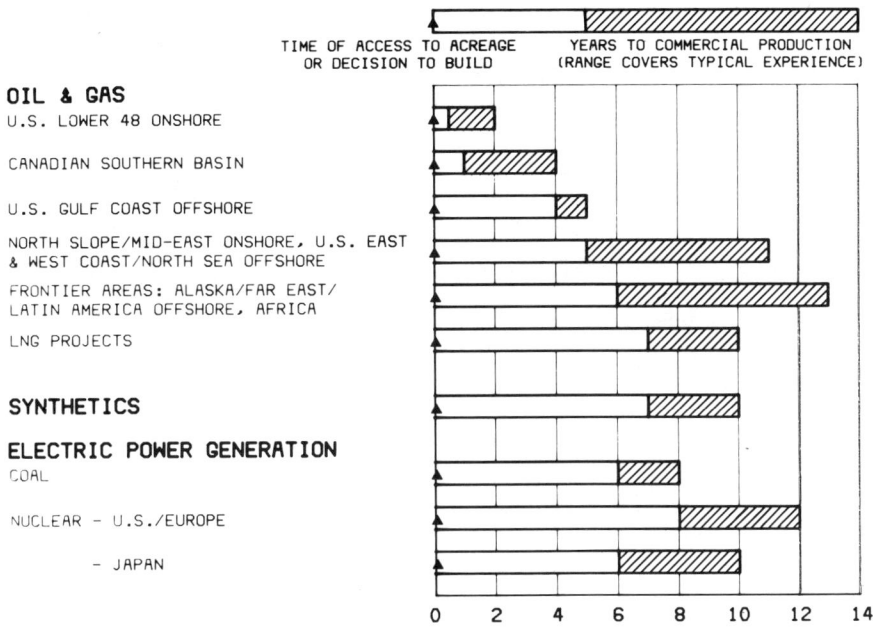

Figure 3-10. Energy Supply Lead Times

French, German, and Italian interests are now constructing the largest—and only—full-scale commercial nuclear breeder reactor in the world.

South Africa expects to cover a third of its petroleum requirements with coal synthetics by 1985.

Uruguay and Brazil will benefit from the construction of the world's largest dam, which will supply nearly 13 gigawatts. Also, other hydroelectric projects are underway in Mexico, Venezuela, and various Central American countries. There are other opportunities to be pursued.

In Canada development of synthetic fuels has been moving forward, for example, with the Syncrude project, in which Exxon's Canadian affiliate is participating. It will produce the equivalent of some 100,000 barrels of oil a day by the end of 1982.

In Queensland, Australia, the Exxon affiliate is participating in initial studies on a major shale project that would ultimately yield the equivalent of 200,000 barrels of oil a day by the mid-1990s. This monumental project will involve the largest earth-moving operation ever attempted. It will be on such a vast scale that it will be equivalent to digging out a Panama Canal every three years.

These are all good signs, and there are many more that could be cited. I hope that public and governmental concerns do not revert to their previous seesaw patterns—rising when things look bad, and going down when crises recede from sight. The world's energy problems, after all, are a thing not of the past, but of the future.

4

Geopolitics of Energy in the Persian Gulf

Edward Teller

Although it is dangerous to make predictions, particularly about the near future, this chapter will in fact address the near future. What will happen in ten years, twenty years, thirty years, may in itself depend on what might happen in one year or four—and there are strong indications that something indeed will happen in the very near future.

The USSR has surrounded the Persian Gulf. South Yemen has been a Soviet satellite for some time and is full of Soviet military equipment. The Soviets helped to bring about the revolution in Iran and thereby destroyed a regime that was far from excellent. However, the former government, as is evident from the increasing instability in that region, certainly was a stabilizing factor. The USSR has moved into Afghanistan, and if one analyzes the reasons, one concludes that they have done so for geographical reasons, further to close the ring around the Persian Gulf. There is discussion that may produce results—I hope it never will—of Yasser Arafat taking over the West Bank.

It is quite clear that there has been preparation. The question is whether this will be followed up by decisive action and, if so, when. If the USSR were to move in the near future, say in the next two to four years, I believe it would be impossible for the United States to put up military resistance in that area. Soviet arms are quantitatively ahead of U.S. arms. The USSR is becoming and in many cases already is qualitatively ahead of the United States as well. This fact is overlooked because technologically the USSR has not performed well in the civilian sector. The situation is further obscured by the fact that whatever intelligence we gather about Soviet accomplishments is more highly classified than intelligence about our own progress. Still, the evidence for Soviet military progress exists.

What may be most important is that domination of the Persian Gulf is an unvarying goal of Soviet policy. Carried over from the time of the czars, it played a prominent role in the Stalin-Hitler pact and continued until 1945, when only the great military superiority of the United States and determined opposition by President Truman prevented the USSR from taking over most of Iraq. Thus the present Soviet moves are an obvious continuation of an old policy.

This policy existed even before the Persian Gulf began to play a decisive role in the world energy situation. Today 45 percent of the oil consumed by the noncommunist world comes from the Persian Gulf. If the USSR controlled the flow of oil, it could bring about a crisis, even an immediate destruction of the NATO alliance. Such a crisis could have a disturbing effect on the U.S. economy and a catastrophic effect, if the USSR wanted to produce one, on the third world. Here are a couple of relevant figures from U.N. statistics.

In the period 1950-1975, the advanced countries increased their per capita energy consumption by 70 percent. In the same period the third world increased its per capita consumption threefold, partly as a consequence of a tremendous population growth. The increase was mainly based on oil—easily transported, easily distributed, easily used in cheap apparatus that is easy to operate. In other words, oil stands at the beginning of industrialization in the third world. For this development small units are best; although in some cases nuclear energy can help, oil is a quicker and better aid. Without oil developing countries will not develop; even worse, without oil they will not eat. Their tremendous population growth, the result of advances in medicine and declining levels of infant mortality, would have long since brought about starvation, were it not for the green revolution. The new crops that have been developed give yields of twice—sometimes even five times—former yields. This farming procedure has been adopted even by primitive populations with remarkable speed because their survival was at stake. But the sad fact is that the new crops will perform less well than the old ones unless there is plenty of irrigation and plenty of energy-rich fertilizer. Now and in the near future, in a practical sense, both of these depend on oil. Therefore, if the USSR commands the Persian Gulf, it can say to many millions of people: Obey or starve.

When will the USSR move? Its preparations appear to be far advanced. In the coming decades—between now and the year 2000—the world's dependence on oil will decrease. Hence the later the USSR moves, the less effect it can produce. Also, with the election of Ronald Reagan as president of the United States, there is no question that the United States will pay more attention to its military preparedness. This may or may not mean more expenditure, but it certainly will mean something that has been lacking—more military research.

Today, any U.S. resistance on the other side of the world, in the face of a strong opponent, is hopeless. The nearest U.S. base is in Diego Garcia, 3,000 miles away—too distant to be effective, but close enough for Soviet bombers to destroy U.S. naval forces. Any boasting about U.S. strength, any drawing of lines is, in my opinion, meaningless. I have strong reasons to believe that President Reagan is more realistic and will not make adventurous moves that are doomed to fail.

The question is, then, what can we do? By we I do not mean the United States; I mean the noncommunist world. The only remedy is to be prepared to exist, for at least a limited time, with greatly reduced oil supplies. I mean that in such a contingency the United States should stop importing oil and, if at all possible—and I think it might be possible—even become for a short time an oil exporter. This cannot be done without reducing energy consumption—in particular, oil consumption. It is not just energy conservation I have in mind, but also drastic measures, such as the decrease in the consumption of gasoline to one-third the present amount. This cannot be done without careful planning. Such planning has not taken place either in Washington or in our big companies. It has not taken place in the other advanced democracies, at least not to a serious extent.

If we can plan, if an aggressive move by the USSR would lead to a closing of ranks in the free world, then that would be the strongest deterrent for the USSR. It would at least reduce their incentive to move. My main wish naturally is that my prediction will not come true. And, of course, I am not predicting military conflict. In fact, military conflict would be so insane in the present situation that the likelihood is overwhelming that it will not occur.

What are the measures we can take? The United States should increase energy production, but that will not suffice because it takes too much time. Still, in the United States alone, if all the nuclear plants that are in an advanced state of construction were given the green light by the Nuclear Regulatory Commission (NRC) (and there are safe plants that should be given the green light), and if all unnecessary restriction on producing plants were lifted, within one year that would reduce U.S. imports of oil by 1 million barrels a day.

Unfortunately, the staffing of the NRC is such that this rapid, necessary, and justified licensing is not now in prospect. I hope it will be possible under the Reagan administration. We must eliminate imports, but using our nuclear-power capacity will save only 1 million barrels out of 7 million, and that only after a delay of one year.

A lot of new gas has been found. To get it to the customer, to drill the wells, to lay the pipelines—all this will take time. The sooner we start production, the better. But if the USSR acts in the near future, this increased production will be too late. All we can accomplish is perhaps to reduce the time of an acute energy crisis from five to ten years to a period of between two to five years. The latter, of course, would be more bearable.

I will not address the question of whether there should be rationing—whether there should be, for instance, a prohibition of air-conditioning in this period. People cannot survive in some parts of the United States without some heat. These are very serious questions, which I will only mention here. I will address two examples—not as recommendations, but as suggested discussion topics. If the United States cuts automobile transportation drastically, then Detroit, which is already in a critical situation, will be in bankruptcy. How severe the blow will be depends on the question of how rapidly Detroit can change, for instance, from making automobiles to manufacturing mopeds.

There is another, similar question. If U.S. automobile production drops drastically, there will be a drop in steel consumption. At the same time, the United States will need steel for oil rigs and pipelines more desperately than ever before. Both planning within industries and planning between industries are necessary if ready responses are needed.

In any economic shock the classical pattern is that the building industry practically stops. This naturally affects the lumber industry. However, this will be precisely when we need wood—not for building houses, but for burning. How quickly we can switch from the one to the other may be crucial. When I refer to contingency plans in the United States, I am talking about questions of this kind.

There should also be contingency plans on an international scale, and there I cannot even begin to suggest how to start the discussion. It is, as I have said, risky to make prophecies about the near future. My main hope is that my prophecy will be wrong. My secondary hope is that if my prophecy happens to be right, the West will be prepared.

5

Annotation on Energy in the Crucial Decade of the 1980s

C. Pierre L. Zaleski

Introduction

Man has always believed that the age he was living in was the "age of problems." Every era has its problems, some of them insurmountable. In our generation the world's energy future has given rise to deep concern—underscored again at the recent World Energy Conference (WEC) held in Munich in September 1980. But perhaps these are only more exaggerations from those pessimistic experts who in 1974 predicted a serious crisis that was, in large measure, overcome. Is the world's medium-term energy future really hopeless? A simple comparison of the legitimate needs of different geopolitical zones and of the resources available to the end of the twentieth century shows the seriousness of the problem.

However, history tells us that man has always solved his problems and that, in fact, man's most brilliant progress has been made in the face of difficulty. In the energy game today, we hold a number of trump cards. We must only decide to play them and to adapt to the new world energy structure that will spring from them. Although the world's energy situation is worrisome, man has the means to avoid a serious crisis, provided all the players—producers and consumers—overcome their semiparalysis and contribute to achieving this new global energy balance.

Energy Needs

Population growth is one of the primary factors governing the increase in energy needs. World population, today around 4.4 billion, will increase by 50 percent in the next two decades and will exceed 6 billion in the year 2000. Over 90 percent of this growth will occur in the developing countries. As for economic growth, there will be a low rate of GNP growth, around 3 percent, in the industrialized countries. But for the non-oil-producing developing countries, where basic needs are still unmet, a growth rate under 5 percent would mean a serious risk of crisis. Thus world energy consumption, now approximately 7 million tons of oil equivalent per year, will rise inexorably to at least 12 million tons a year in 2000. Developing countries will absorb 35 percent of this growth. To satisfy this rate of growth—an average of 2.8 percent per year—while maintaining the same relative shares of different fuels (oil, coal, gas, and so on) must be regarded as utopian. Not only are there limits to fossil-fuel reserves and productive capacity, but also the world cannot tolerate the economic consequences of the unavoidable rise in costs.

Energy Supply

In 1980 some 50 percent of world primary energy needs were supplied by oil, corresponding to annual production of 3.2 billion tonnes. In the year 2000, even though annual world oil production will probably reach 3.5-4 billion tonnes, this will represent a maximum of 30 percent of world needs at that time. Natural gas can be expected to supply no more than 16 percent of world energy needs in 2000 as a result of limited resources, geographical distribution, and transportation. The world energy crisis is above all a petroleum crisis, resulting not so much from limited reserves as from the limits to production and skyrocketing prices. Do we *now* have the means to free ourselves from our current dependence, or must we be resigned to repeated price increases and supply crisis?

New Energy-Supply Scenarios

In the near term there are two ways of reducing the importance of oil in the supply-demand balance: (1) conservation, and (2) substitution of new energy sources for oil. The first demands some caveats. We must take care not to attribute the consequences of the current economic slowdown to conservation. Moreover, the conservation successes realized since 1974 must be seen as a unique phenomenon, made possible by the waste of previous years but not repeatable. The next increments of conservation will require an increasing investment effort. Energy must not be a brake on economic growth; rather, it is a rational use of energy that must prevail.

Renewable Energy Sources

Solar, geothermal, and biomass can all play a role in the world's long-term energy supply. But it would be dangerous to pretend that simply pouring more resources into development of these sources can provide a quick solution to our quantitative supply problems through renewable energy sources. These sources will have a predominantly local impact in this century; including hydroelectricity, they may represent 5 or 10 percent of total world consumption in the year 2000. The major near-term effort should be devoted to reducing their investment cost so that they may play more than a limited role.

Coal can be a significant help, especially for countries like the United States that have large resources. But the growth rates now being predicted for coal are very ambitious. In 1980 coal supplied around 25 percent of world primary energy consumption; international coal trade represented only 7 percent of this production (as opposed to 50 percent for oil). To increase this trade fivefold or even tenfold (for steam coal) by the year 2000—as the recent world coal study projects—implies gigantic efforts of production, port development, and transportation, not to mention the environmental problems that will result. As for synthetic fuels from coal, they can provide one solution but only in countries with large coal resources. Here again, time and investment factors must not be underestimated. The price of a barrel of synthetic oil is estimated today at between $40 and $60; the investment varies between $50,000 and $100,000 for each barrel per day. In the year 2000 coal should be able to supply 30 percent of world energy consumption, with coal-based synfuels representing only a small share.

Nuclear energy is the second major alternative to oil. It is already widely industrialized, rapidly available, economically competitive, and capable of an important contribution to solving the world energy problem. In 1980 nuclear generated more than 600 trillion kilowatt-hours in the noncommunist world, with an installed capacity of 133 gigawatts. This is the equivalent of 130 million tons of oil—the entire yearly production of Venezuela. More important, nuclear power, including fast breeder reactors, offers long-term security of supply. With an aggressive development policy nuclear could supply 15 percent of world primary energy needs at the end of the twentieth century. Because of its high-technology nature, nuclear is the logical replacement energy for the industrialized countries, in some of which it may supply 30 percent of energy needs in 2000. This assumes parallel development of electricity use, particularly in industry, which, although it has problems, also offers the advantages of flexibility.

Hence there are potential means of freeing ourselves from oil. What policy is needed to put them to work? We need all the resources available; this message has been delivered but not translated into action. Given the inertia of the energy situation, the world's political and economic situation in the medium term depends on decisions that must be taken today. The energy situation leaves no room for hesitation but requires quick action, determination, and perseverance. The energy crisis is serious and will not go away by itself. The resources and human capacity are there. The only question—and it is a hard one—is the political will.

6 Toward a New Order for the Global Energy System

Wolfgang Sassin

Introduction

For almost a decade we have had to live with what we call the energy problem. This problem has sometimes taken the form of supply embargoes and in other periods has manifested itself in dramatic increases in the price of crude oil, which determines the price of all other energy sources. The nuclear-energy dispute has further complicated the problem. In recent years more general economic difficulties have emerged that are directly and indirectly linked to the energy problem. These include a decline in the potential for economic growth, restrictions in international trade, and serious imbalances in the international monetary system. Despite a number of attempts to overcome such difficulties, no convincing solutions have yet been formulated, let alone put into practice. As things now stand, no end to the so-called energy problem is in sight.

We must therefore investigate the root causes of this phenomenon before it becomes totally impossible to control its future development. Such an analysis must be based above all on a painstaking scrutiny of the decision-making principles that bind technical and economic measures relating to energy into a system. One possible way of identifying inherent contradictions within this energy system and the crucial developments to which they give rise is to extrapolate existing trends. The aim is not so much to arrive at the most probable forecast of future developments as to throw light on the underlying assumptions and preferred options. The further into the future a projection of this kind is taken, the more it tells us about inherent conflicts that now remain obscure.

During the seven-year period 1973-1980, the International Institute for Applied Systems Analysis (IIASA) conducted a global systems analysis of the energy problem, considering prospective demand and modes of supply for the fifty years from 1980 to 2030. The central findings of this work have been published by the Energy Systems Program Group of IIASA (Wolf Häfele, program leader), vol. 1, *Energy in a Finite World: Paths to a Sustainable Future*, by Jeanne Anderer with Alan McDonald and Nebojsa Nakicenovic; vol. 2, *Energy in a Finite World: A Global Systems Analysis* (Cambridge, Mass.: Ballinger, 1981).

Possibilities of Technical/Economic Equilibrium between Energy Demand and Energy Supply

As we are concerned here not with formulating an energy strategy but with gaining a better understanding of the energy problem, we can take any of the numerous long-term projections as a basis for quantitative considerations. The only important factor is to ensure that we select a comprehensive projection—one that includes all the economic regions of the world. Otherwise there is a danger that the contradictions we seek to identify will tacitly be ascribed to those areas or sectors that have been excluded from detailed analysis.

The scenarios published by the International Institute for Applied Systems Analysis (IIASA) in Laxenburg with a time scale of fifty years are a suitable projection of possible developments in the global energy system.[1] Detailed economic-growth projections have been drawn up for each of the seven world regions shown in figure 6-1, which combine estimates of demographic, technological, and economic trends. The calculations of regional energy demand obtained were then compared with estimates, also on a regional basis, of total utilizable resources of energy and the rates of expansion of sophisticated energy-supply technologies.

In its simplest terms, the energy problem initially presented the analysts with the dilemma of the fundamental irreconcilability between high demand for energy as a result of high rates of economic growth in all regions and the global supply of energy, which was too low even on the basis of optimistic estimates. This gap has finally been overcome by means of an extremely complex multistage iterative process, which in the final analysis consists of dividing and relocating the problem at other levels.

The major assumptions and principles that finally led to a mathematical balance—that is, a systems-analysis solution of the energy problem—are, briefly:

1. The IIASA scenarios assume a stable political development without major military confrontations or fundamental social upheavals.
2. The energy resources of each region are available at cost. All regions can make use of energy technologies, including their applications, on the same terms. Apart from conventional oil, energy is exported and imported between the regions at cost.
3. Each region will expand its energy-supply system over a period of fifty years, taking into account its demand and resource situation, and considering minimal overall costs.
4. An oil-production ceiling of 33 million barrels per day was assumed for the countries in region VI. This quantity covers domestic demand and determines the maximum level of exports.

Toward a New Order 49

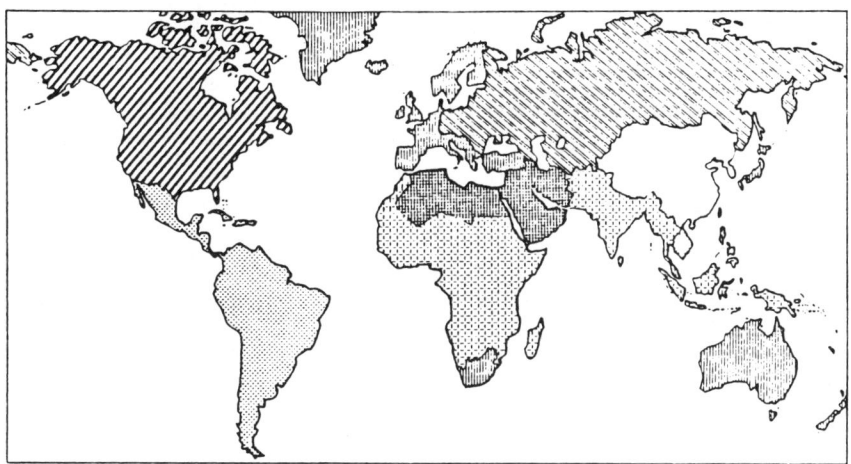

	Region I	(NA) North America
	Region II	(SU/EE) Soviet Union and Eastern Europe
	Region III	(WE/JANZ) Western Europe, Japan, Australia, New Zealand, S. Africa, and Israel
	Region IV	(LA) Latin America
	Region V	(Af/SEA) Africa (except Northern Africa and S. Africa), South and Southeast Asia
	Region VI	(ME/NAf) Middle East and Northern Africa
	Region VII	(C/CPA) China and Centrally Planned Asian Economies

Figure 6-1. The Seven World Regions Analyzed in the Studies of the International Institute for Applied System Analysis, Laxenburg, Austria

In such a projection these basic assumptions, in conjunction with other norms that guarantee a technically efficient use of the individual processing and application techniques, provide an estimate of the minimum technological and economic effort necessary to satisfy a given consumer demand for energy. In practice this minimum is always exceeded.

Despite such extreme assumptions it nonetheless proved necessary to reduce even further the original modest estimates of growth rates.[2] In the absence of such an adjustment, with all its implications for the overall development process, the rates of development of most alternative sources of energy—such as coal, nuclear energy, and solar technology—would have been completely unrealistic.

This crisis indicated by the computer model discouraged the IIASA from developing an average or most probable projection. High and low scenarios for economic growth were produced to cover a wide range of conceivable situations in which energy supply and demand were in a state of equilibrium.

Table 6-1 and figure 6-2 provide data showing the consequences of such a gradual solution of the global energy problem. The rates of economic growth achieved, particularly in the low scenario, raise the question of political stability in the developing regions. It is important when considering these figures to stress the importance of the basic assumption of free access to global energy resources and optimum technology. The relatively high rates of energy saving (figure 6-2) and the rates of expansion of fossil fuels, nuclear energy, and renewable sources of energy (table 6-2) pose additional problems.

Prerequisites for the Realization of the IIASA Scenarios

This section is less concerned with the probability of the IIASA global scenarios being realized in the future than with identifying some of the quantifiable prerequisites that would permit these systems-analysis solutions of the energy problem to be attained. We have concentrated on the monetary aspects.

Table 6-1
Economic Projections Influenced by the Energy Problem: The IIASA High and Low Scenarios

Region	Historical Growth Rate of Per Capita GDP (%/yr) 1950-1975	GDP Per Capita (Dollars) 1975	Projected Growth Rate of Per Capita GDP (%/yr)			
			High Scenario		Low Scenario	
			1975-2000	2000-2030	1975-2000	2000-2030
I (NA)	1.9	7,046	2.9	1.8	1.7	0.7
II (SU/EE)	6.7	2,562	3.6	3.2	3.1	1.9
III (WE/JANZ)	4.0	4,259	3.0	1.8	1.7	0.9
IV (LA)	2.9	1,066	3.0	2.4	1.6	1.9
V (Af/SEA)	2.5	239	2.8	2.4	1.7	1.4
VI (ME/NAf)	5.7	1,429	3.8	2.8	2.4	1.2
VII (C/CPA)	5.1	352	2.8	2.4	1.6	1.4

Note: All growth rates are average annual growth rates (rounded) over the time period shown; actual projections have decreasing growth rates.

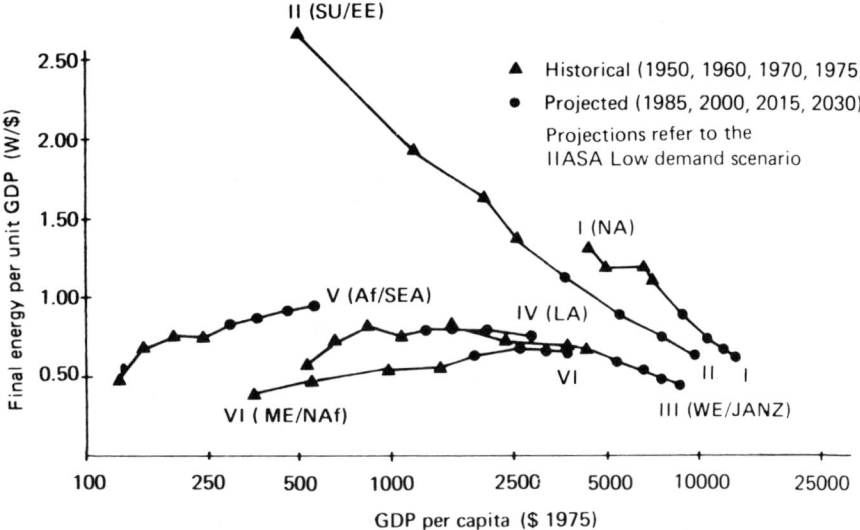

Figure 6-2. Macroeconomic Energy Efficiency as a Function of Economic-Activity Level

The scarcity of high-grade energy reserves means that lower-quality fossil resources must be developed. This involves not only higher operating costs but in particular a far higher expenditure of capital per unit of utilizable energy obtained. Investment costs not only in the mineral-oil and natural-gas sectors, but also in the coal and nuclear-energy sectors, have risen drastically since the beginning of the 1970s, and this trend will continue. The inclusion of conventional energy reserves that had previously been regarded as uneconomical and the exploitation of unconventional

Table 6-2
Two Supply Scenarios, Global Primary Energy, 1975-2030
(TW)

Primary Source	1975	High Scenario		Low Scenario	
		2000	2030	2000	2030
Oil	3.62	5.89	6.83	4.75	5.02
Gas	1.51	3.11	5.97	2.53	3.47
Coal	2.26	4.95	11.98	3.93	6.45
Nuclear 1	0.12	1.70	3.21	1.27	1.89
Nuclear 2	0	0.04	4.88	0.02	3.28
Hydro	0.50	0.83	1.46	0.83	1.46
Solar	0	0.10	0.49	0.09	0.30
Other	0.21	0.22	0.81	0.17	0.52
Total	8.21	16.84	35.65	13.59	22.39

sources have transformed the problem of scarcity in the IIASA scenarios to one of expense. Figure 6-3 shows the annual investments needed to provide secondary energy as a proportion of projected regional gross domestic product (GDP). These investments include the cost of exploration and exploitation of oil and gas fields; of underground and surface mining of coal, oil shale, tar-sands and uranium; of power stations, gasification and liquefaction plants, and plants to produce capital goods needed in energy production. They do not include transport costs, the costs of distributing energy to consumers, or investments for energy-saving equipment.[3]

In the high scenario, energy investments until the year 2030 for supply systems that already take into account minimum overall costs in the seven world regions amount to U.S. $50,000 billion at 1975 prices (approximately $30,000 billion in the low scenario). To illustrate the scale of these investments, consider that world GNP in 1975 amounted to approximately U.S. $5,000 billion. It is important to note the steadily rising proportion that energy investments claim from the economy of the developing countries in the IIASA scenarios (figure 6-3). In the period under consideration, this triples, whereas in the industrialized countries it rises only to approximately 1.5 times the present proportion. At first these ratios appear difficult but by no means impossible. However, a confident judgment at this level of analysis would be possible only if the investments required to overcome other problems, such as shortage of food, housing, or employment, had been quantified.

Viewed in macroeconomic terms, the energy problem thus appears soluble on the basis of the foregoing premises by means of free access to global energy resources. At the same time this involves a steady rise in expenditure,

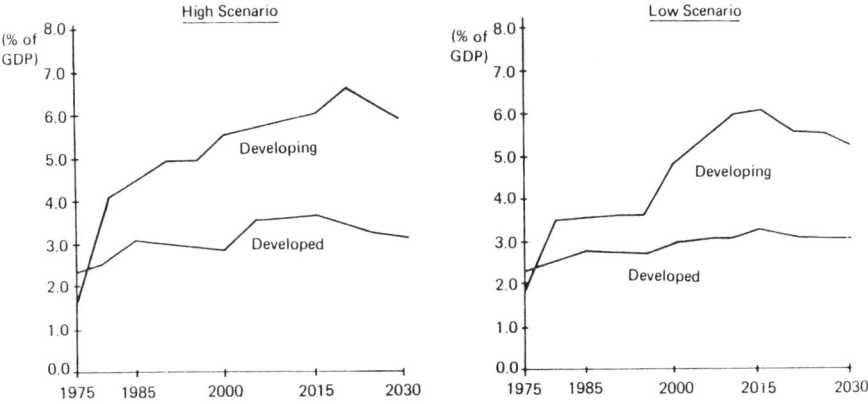

Figure 6-3. Calculated Direct and Indirect Investments to Build Up the Energy-Supply Systems of Two IIASA Scenarios

Toward a New Order

which in turn means rising energy prices. The expenditure consists of capital-servicing costs for investments and the running costs of the energy infrastructure in each case. As this infrastructure changes in the course of time, and moreover is not the same in every region, these costs for individual sources of energy in each region will also vary and can be calculated from the IIASA scenarios. The underlying assumptions that energy imports and exports take place to a certain extent between the regions means that costs alone do not determine prices, and thus the volume of trade, but that these also depend on the possibilities of substitution among primary energy sources. The cost to any given regional energy system of producing one additional unit of energy from other, as yet nondepleted, sources is expressed as a marginal cost. These marginal costs vary in the course of time. If one assumes that energy prices in the long term will not contain a political component, then these marginal costs represent average maximum market prices.

Figure 6-4 shows the marginal costs for the most important group of international commodities—liquid fuels. The worldwide average marginal cost rises in both the high and the low IIASA scenarios by a factor of

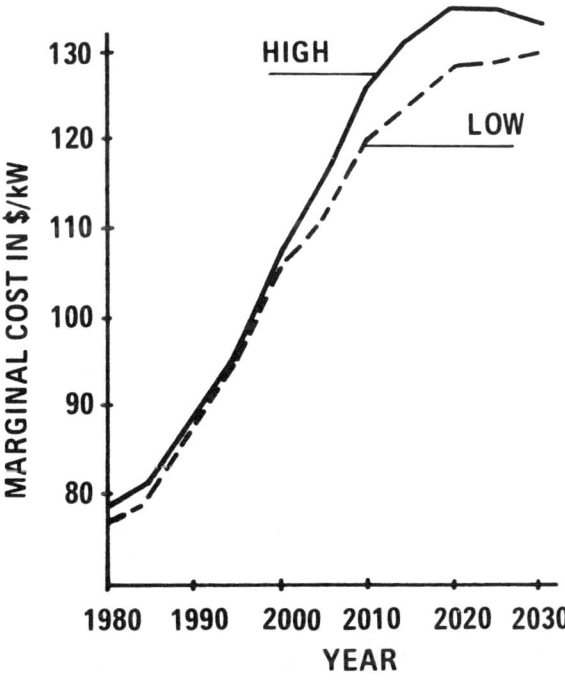

Figure 6-4. Time-Sliding Worldwide Average Values of Liquid-Fuel Price, High and Low Scenarios, 1980-2030 (Excluding Region VI)

approximately 2 between 1975 and 2030.[4] There are, of course, major differences between regional marginal costs and the worldwide average. In the final analysis these differences simply reflect the relative wealth in readily available energy resources of the various world regions. These differences cannot be entirely offset by technological progress, given the assumption of free access to new energy technology.

Within the logic of the technical economic ideal solutions, therefore, some world regions will remain dependent on energy imports even in the long term. This applies particularly to the industrialized countries in region III, which possess few domestic resources. As demand for energy rises, the countries in Central Africa and Southeast Asia are also likely to become dependent in this way. Figure 6-5 illustrates for both IIASA scenarios the crucial relationship that will develop as a result of competing demand between region III and region V for oil from the Middle East and Latin America.

When one considers the political problems in the Middle East, the balance-of-payments difficulties that regions III and V are already experiencing and the slow progress of region I in developing its own resources, then figure 6-5 underlines the scale of problem-solving capacity needed outside the technical/economic sector for the IIASA scenarios to be realized.

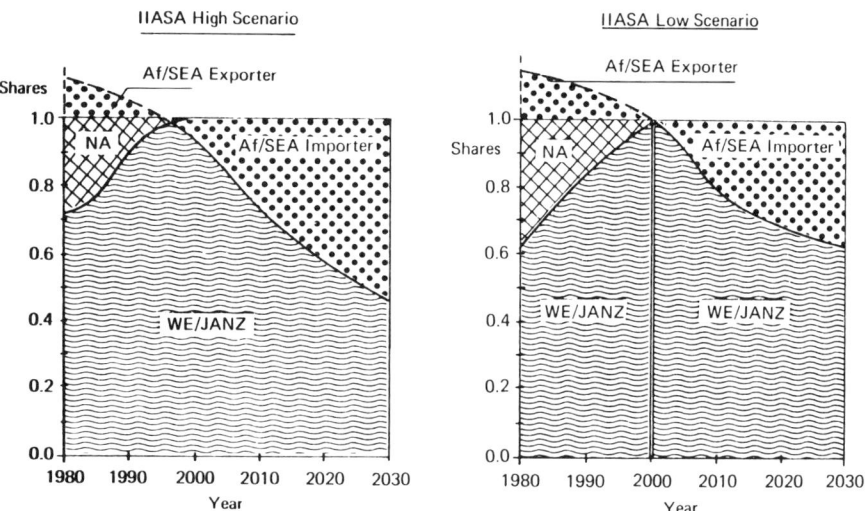

Figure 6-5. Allocation of Energy Exports from Resource-Rich Developing Regions, Middle East/North Africa and Latin America

Conflict between Macroeconomic and Microeconomic Optimization

Even assuming that a satisfactory solution to the more general political disputes between the regions can be found so that energy no longer needs to be used as a means of diverting industrial productivity or achieving political goals (and this is generally taken for granted in the IIASA scenarios), the energy sector still confronts a major dilemma. We can illustrate this dilemma by asking whether the energy sector at either global or regional level will be able to finance the necessary restructuring and absolutely essential expansion from its own resources. The quantitative IIASA scenarios go some way toward answering this question, too.

Table 6-3 shows annual turnover from liquid fuels in relation to annual rates of investment to expand and restructure the global energy system. Turnover is expressed in the marginal prices from figure 6-4. Liquid fuels account for just under half of worldwide demand for secondary energy. Total turnover in terms of value is likely to be roughly twice as high if one allows for the somewhat lower prices of solid fuels and the higher prices that can be charged for electricity. Even by 1990 worldwide total annual energy investments are as high as the postulated turnover from liquid fuels. They remain at this level, with minor fluctuations, until 2030. If one considers that average energy costs at present, which consist of capital and running costs, can in an optimum system amount to only a little less than the marginal prices for oil as the main source of energy, then with the prices postulated here the energy sector cannot achieve the surplus needed to finance its own long-term investment program.

Although it would be possible to resolve the microeconomic dilemma of financing an energy infrastructure that made macroeconomic sense purely in accounting terms by charging significantly higher prices for energy than

Table 6-3
Liquid Fuels and Monetary Flows in the Global Energy System

Year	Total Liquid Fuels High/Low TWyrs/yr	Marginal Systems Price U.S. (1975)$/kW (per BOE)	Annual Turnover of Liquid Fuel Ind. at Marginal Prices 10^{12} U.S. (1975)$	Investments Total Annual for Global Energy Supply System 10^{12} U.S. (1975)$
1975	3.4	70(14)	0.24	0.16
2000	5.5/4.4	105(21)	0.58/0.46	0.62/0.47
2030	10.4/6.7	135(27)	1.40/0.90	1.50/0.87

those in table 6-3 or figure 6-4, this would result in a situation that militated against a localized cost minimization because areas with surpluses would be forced to compensate for losses from other energy activities by transfer payments. The geographical distribution of the more readily available energy resources means that this would involve international transfer payments or preferential prices.

Thus the demand for microeconomic financing of the infrastructures needed by both national economies and the world economy leads inevitably to a demand for market organization and international conventions. Even if the microeconomic financing deficits were to be made up by the taxpayer in the form of subsidies or deficits of nationalized companies, there would still be the question of the criteria that would guarantee secure long-term development of the global energy system.

Before drawing conclusions from this new situation, we should first consider an obvious objection. It could be argued that market forces are able to overcome even the most difficult structural crises. The energy sector in particular has an impressive record of successes in changing over from wood to coal, from coal to oil and gas, and—at least to some extent—to expensive nuclear energy. But this process of constant economic innovation for over a century was possible only as long as technological progress provided access to sources of energy at steadily decreasing cost. The result of this technological progress was that the surplus that the market yielded always went to the innovator. Microeconomic cost minimization and market penetration led to overall systems economically efficient from the macroeconomic point of view. This system of convergent decision-making criteria was bound to lose its inner stability once technological progress was no longer able to offset a decline in the quality of natural resources by increasing productivity within a given sector. The global energy system appears to have reached this turning point during the 1970s. If present estimates of long-term increases in the cost of producing and processing energy are correct, then the microeconomic stimulus to innovate, or even to invest generally in energy plants, is lacking—a stimulus that has been the driving force behind the energy system in the past century. Figure 6-6 attempts to show how an incentive gap, which is now unfortunately a real possibility, can come into being. The roots of the present energy problem, therefore, lie less in the imperfections of economic systems than in the limits to technological progress that are beginning to become apparent.

Conclusions

If it is true that over the next few decades it will prove impossible to counteract a steady increase in average energy costs by means of technological progress, then energy markets will have to be stabilized by ex-

Toward a New Order

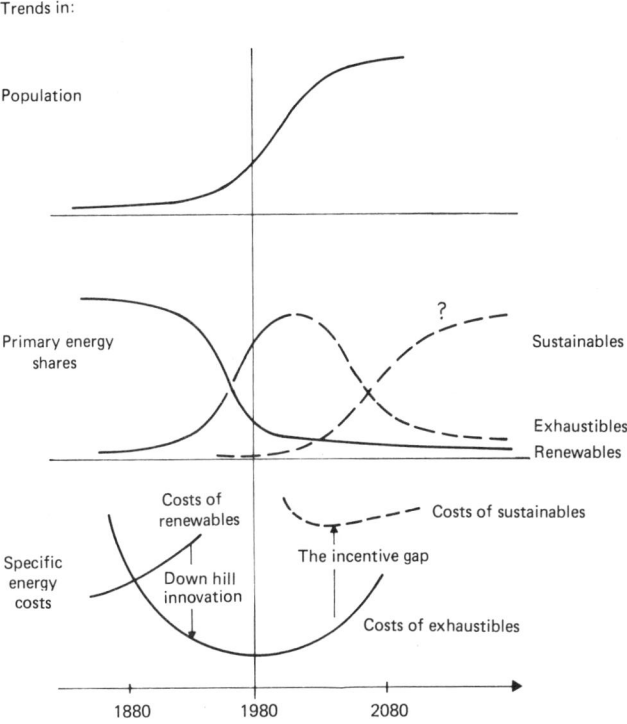

Figure 6-6. Incentive Gap for a Transition from Exhaustible Fossil Energy Resources to Sustainable Supplies from Nuclear Breeding and/or Direct Uses of Solar Radiation

ternal influences. It is impossible to predict how this can be done until we have first carefully studied the control mechanisms available and their repercussions on general economic activity, particularly the international monetary system.

Nonetheless, we can already make a few basic comments on the direction in which such controls should lead. They should compensate those firms that innovate to build up a macroeconomically optimal energy supply system for regional disadvantages and all operating losses incurred in periods of technical depreciation. In addition, steps should be taken to ensure that existing global energy resources are used in an economically optimal manner and to aid the transition to a global energy system that will be viable in the long term. In other words, as far as possible the energy system must be self-financing on a worldwide basis. At all events steps should be taken to prevent existing problems on the energy markets leading to a resubstitution via scarcity rents of capital and labor for energy. Ultimately this would slow down the global process of economic development far more than it would provide benefits in terms of short-range advantages from scarcity rents of energy.

From this point of view, the recycling of petrodollars into the energy system of those countries that are poor in resources is an appropriate and clearly urgent measure. But it requires not only agreement on the long-term goals of economic development but also coordination of the operational targets leading toward a new order for the global energy system.

One of the most important problems to be overcome in the 1980s undoubtedly is that of transferring monetary values from the overall economy of the individual industrialized countries to those areas of the national energy economy that need support in order to supply a sufficient quantity of competitive energy and to those consumer groups that need the support to save energy without incurring individual disadvantages. It should also be noted that for the national economy a shortage of one barrel causes today an approximately ten times higher loss in the GDP than the economic equivalent of its production or saving. What support will be needed for the rapid utilization of alternative energy sources will depend on the particular resource situation of the individual countries. It seems that the sharp differences in the competitive situation among the individual energy sectors of a country can be balanced in view of the low burden of the overall national economies that would organize this kind of transfer payment for their energy sectors. Only if the disadvantages caused by varying resource situations are distributed within larger economic groups of a nation can a stabilization of the energy problem be expected.

Notes

1. W. Häfele, "A Global and Long-Range Picture of Energy Development," *Science* 209, no. 4452 (Centennial Issue, July 1980): 174-182; W. Sassin, "Energy," *Scientific American* 243, no. 3 (September 1980): 106-117.

2. W. Häfele and W. Sassin, "Resources and Endowments: An Outline of Future Energy Systems," in P.W. Hemily and M.N. Özdas, eds., *Science and Future Choice* (Oxford: Clarendon Press, 1979).

3. Yu. Kononov and A. Por, "The Economic Impact Model," RR-79-08. International Institute for Applied Systems Analysis, Laxenburg, Austria, forthcoming; L. Bauer, W. Häfele, and H. Rogner, "Energy Strategies and Capital Requirements," Paper submitted to Eleventh World Energy Conference, Munich (1980).

4. A. Papin, "Domestic Liquid Fuel Prices by Seven World Regions," Forthcoming research report, International Institute for Applied Systems Analysis, Laxenburg, Austria.

7

International Aspects of U.S. Energy Choices

Sam. H. Schurr

This chapter deals with the findings of the study I did for Resources for the Future—since published as the book *Energy in America's Future*—dealing with the United States' energy choices. Although the orientation of that study was essentially domestic, it is self-evident that how the United States deals with its domestic energy situation inevitably carries strong implications for the rest of the world. The sheer weight of the United States in international energy markets would alone give domestic decisions affecting oil imports an extraordinary degree of international importance. Equally important is the fact that the productive strength of this country can be crucial in determining the future of the entire world.

Heavy reliance on insecure oil imports is at the heart of the immediate dangers with which U.S. energy policy must deal. But as great as these immediate dangers are, they do not fully define the energy problems that the United States or, for that matter, the world faces. At the top of today's energy-policy agenda must also be the related problem of managing a long-run energy transition: away from resources such as petroleum and natural gas that appear grossly inadequate for the future, and toward supply sources such as coal, nuclear fuels, and renewables, for which resources are sufficient to meet the long-term needs of a growing world economy.

For those problems arising out of excessive reliance on insecure imported oil, we are now in a time of maximum peril. This situation calls for action along several lines. Dependence on oil imports and the associated vulnerability to supply interruptions must be sharply reduced through conservation and interfuel substitution. Further, plans and programs must be in place for managing the supply emergencies that would be caused by major import disruptions if these should occur—as very likely they will. We are moving ahead in all these areas; but on some vital requirements, such as filling the strategic petroleum reserve, progress so far has been embarrassingly and dangerously slow.

These immediate problems pose an enormous challenge. However, as though this were not enough to deal with, we must also begin to make decisions and take actions that will lay the foundations for the long-term energy future. Energy solutions, particularly those requiring technological development and capital turnover, have long lead times. To speak in terms of decades does not exaggerate the time sequences involved. Thus on these matters near-term decisions are required in order to meet long-term objectives.

In making these decisions for the more distant future, it is essential to take full advantage of the available potentials. If there is a bright side to all of this, it is that potentials for the long run are quite promising, in contrast to the tightly constrained situation for the immediate future.

The most fundamental decision about the future that will have to be made is the choice between accepting and accommodating to chronic energy scarcity (the road we are now on) *or* creating the conditions that will ensure energy abundance over the longer term. It is vital to understand that there *is* a choice between these two alternatives, because which of the two becomes the real future will not be a consequence of the playing out of predetermined trends, but instead will be determined by the policies we pursue. Policies in turn will be shaped by what we believe the facts to be on such underlying questions as natural-resource potentials, the prospects for supply and demand technologies, the environmental impacts of energy production and use, and the energy requirements for future economic growth and well-being.

The main purpose of *Energy in America's Future* was to assemble and analyze the facts on these subjects and to interpret their implications for national energy policy. I will now summarize some of the findings, beginning with future energy consumption.

We made a range of consumption estimates for the United States to the year 2000, based on a detailed analysis by sectors of use. Our midrange calculation, to which we attach the most credibility, comes to about 115 quads, compared with a 1976 level of about 75 quads (1 quad = 10^{15} Btu, roughly equivalent to the energy content released by burning 500,000 barrels of oil per day for one year). This implies an annual average energy growth rate of 1.8 percent to achieve an assumed GNP growth of 3.2 percent per year (the latter based, as is customary, on labor-force and productivity potentials). In other words, energy consumption will likely grow at a substantially slower rate than the overall national output; conservation will in fact have a significant impact in holding down consumption.

Automobiles and residential heating, which accounted for about one-quarter of all the energy consumed in the United States in 1976, illustrate the difference conservation is expected to make. We estimated that by the end of the century there would be over 150 million vehicles, up from 110 million now. But average fuel efficiency should about double during this period, thus allowing the absolute level of fuel consumption for automobiles to decline by 30 percent. Housing units are also expected to increase substantially, from about 70 million in 1976 to 115 million in 2000. But with improvements in the housing shell and increased efficiency in the heating system, total energy used in home heating should also drop.

Residential heating and automobiles are, of course, two familiar categories of use that are referred to frequently to demonstrate the favorable potentials for conservation. Our quantitative estimates support such expectations. Even so, after an examination of all sectors of use in considerable detail, our estimates also show that the *overall* growth of the total economy, including the all-important productive sectors such as industry, services, and commercial transportation, will require a 40-quad increase in the total level of energy consumption in the year 2000 compared with 1976. Such an increase implies cumulative consumption over the intervening quarter century of around 2,500 quads. This is a large amount—equivalent, for example, to considerably more than *all* the estimated remaining recoverable domestic resources of conventional oil and natural gas, conservatively estimated. Thus a substantial expansion of energy supplies will be needed, and it cannot come from petroleum and natural gas, as has been true in the past.

The prediction of an urgent need for supply expansion to meet consumption growth runs counter to a school of thought that has gained considerable attention recently, namely, that energy conservation, if pressed with enough vigor, could virtually eliminate the need for efforts to expand traditional sources of energy supply. In our projections, we tried to measure only those reductions in energy use that could be brought about in a cost-effective way. We were concerned about the possible negative repercussions on economic growth and the productive efficiency of the national economy that could result from a single-minded pursuit of fuel-efficiency targets to the neglect of other objectives.

Of particular relevance is the relationship between the use of energy and the productivity of labor and capital employed in processes of production. In the past there has been a strong connection between the quantity and form of energy used and the growth of productive efficiency. U.S. economic history is eloquent on this score. If in the future we substitute away from energy in the productive sectors of the economy because of its higher relative price, the United States could easily wind up with losses in productivity as a result. When it is cost effective to do so, substitutions away from energy will take place automatically; but we should not look on all such substitutions with complacency because the total output of goods and services could well be lower than if energy were used more freely in response to lower costs. Pushing reductions in energy used in production processes even further—by imposing taxes on its use or through other policy instruments—could compound the problem of productivity losses. Unquestionably, fuel efficiency in productive operations—in addition to the savings to be expected in the household sector—is an important objective. But it should be pursued with careful attention to its possible detrimental effects on the overall efficiency of the national economy, not just its positive effects on fuel efficiency. By the same token, selecting and developing those supply technologies that will keep energy costs as low as possible should be an important national objective if energy use is to resume its traditional role in enhancing the efficiency of labor and capital used in productive operations.

This is a feasible objective because the potentials for supply are really quite favorable. Conservative estimates show that there is in the United States a reasonably assured domestic-resource total for all mineral fuels used in today's commercial technologies—consisting mainly of coal and uranium—ranging from about 40,000 to 110,000 quads (depending on whether the energy content of uranium is measured with or without "breeding"). This total may be compared with an estimated cumulative total consumption in the last quarter of this century of about 2,500 quads, according to the middle-range projections just cited. Clearly there are enough mineral-fuel resources, conservatively estimated, to carry the United States for a long time. However, these are essentially resources of coal and uranium, not petroleum and natural gas. The crucial issue concerns the future costs of producing energy in the forms desired—liquids, gases, and electricity—from these abundant resources, and their noneconomic characteristics, such as health, safety, and environmental impacts, that play such a large part in public acceptance.

I will pass over the precise cost calculations that, for example, establish small differences in order to make a case favorable to coal or nuclear power in specific regions at the present time. Arguments on such details, as important as they are to current business decisions, have the unfortunate side-effect of diverting attention from the major considerations that should determine broad national strategy for the long term. The most significant element here is that increasingly our basic energy sources will drop the cost characteristics of depletable extractive commodities and will take on those of reproducible manufactured products. This would be particularly true of nuclear energy produced in breeder reactors, for which the costs of raw uranium could become a vanishingly small element per unit of energy production. This characteristic would be less pronounced in processes based on coal and oil shale, but even in these processes the cost of useful oil and gas produced will over time be determined less by rising extractive costs than would have been the case for the petroleum and natural gas replaced.

After technologies of this type mature, their long-run cost profiles are likely to take the general form of a plateau at a real level that we estimate at no more than about twice our base-year prices for liquids and gases and less than that in the case of electricity—rather than a series of sharp steps upward. In other words, a virtual long-run ceiling on the real costs of energy may be possible with supply technologies that are within reach. Thus long-

International Aspects of U.S. Energy Choices

term cost behavior should depart markedly from common expectations of continuously and sharply rising costs for industries based on depletable resources. This fact is crucial to the long-term outlook and would, of course, be rendered even more favorable by commercial success with solar energy and other renewable resources. We concluded, therefore, that the long-term energy supply outlook is potentially favorable not just in quantity terms, but also in cost terms, compared with the dire predictions that are currently so widespread. In other words, cost rises can be contained, and supply abundance can be assured. This would be a change of major historical significance for the United States and the world, compared with the view being so widely advertised by today's neo-Malthusian prophets.

The positive international impacts of U.S. energy decisions are nowhere more weighty than in creating conditions of domestic energy supply that can support a dynamic, rapidly growing, productively efficient U.S. economy. There is sometimes a tendency to decry domestic supply expansion as exacerbating the divisions that already exist in a world characterized by the rich/poor, North/South international split among countries.

However, all countries contain a mixture of rich and poor people. People who consider themselves poor in the United States may not be at the desperate level of poverty found in many of the developing countries, but we must not forget that they regard themselves—*within our society*—as seriously disadvantaged. The social and economic stability of the United States and other developed countries depends, in the first instance, on being able to cope with their own internal problems of income distribution. It is worth remembering that the social and economic stability of a country like the United States, and its ability to discharge international obligations, is vitally important to the stability of the entire world.

There is, of course, no argument with the proposition that world stability depends on improvements in the economic conditions of developing countries, but it is not all obvious how such a result can best be accomplished. In particular, it is not clear that using less energy in the developed countries in order to leave more for the developing world, as though there were so limited and energy-resource base in the world as to require international rationing of its use, is a sensible approach. To the extent that there may be such resource limits, they apply particularly to petroleum and natural gas. Since there is likely to be a continuning squeeze on these energy sources, and since oil in particular appears to be the most desirable energy form for many uses in developing countries, this is all the more reason that the United States, which has the potential for large-scale development of substitute sources over time, should feel both a domestic and an international obligation to do so.

One further point must also be made. If one takes seriously the need to accelerate the growth of the developing countries, this argues for maximum international efficiency in the production of all goods in order to make a greater surplus over and above current consumption available for world developmental purposes. This requirement appears also to argue for rapid economic growth (and associated growth in energy use) in the developed parts of the world in order to enlarge the dimensions of the pie on which all parts of the world can draw in order for world development to proceed at the fastest possible rate.

With such favorable potentials apparently within reach, why do coherent national policies to foster energy expansion remain so elusive? One reason is that there is a lack of appreciation of the real promise of long-term supply potentials. Much needs to be done to create greater awareness of the facts underlying the positive view of long-term supply prospects. Another formidable barrier to supply policies lies in public fears about the environmental, health, and safety aspects of all energy technologies—except perhaps for solar and other renewables.

Here, too, our findings show far brighter potentials than are generally believed to exist. In significant ways, new supply technologies based on coal and nuclear resources are actually less damaging to health, safety, and the environment than are those they would replace. Nuclear power, in its routine operations, causes fewer fatalities and illnesses, and less damage to land, than the chain of supply required in today's technologies to get electricity from coal. New technologies for using coal directly and for converting coal to liquids and gases should be less polluting than the old coal-burning processes they would replace. These technologies in effect convert dirty coal into much cleaner liquid and gaseous forms to be used for combustion. Particular solar technologies do appear to be less damaging than either coal or nuclear; and this may justify a subsidy, representing their social benefits, that would help to offset their monetary-cost disadvantages.

This hopeful side needs to be understood. But there are also persistent concerns about potentially catastrophic threats from supply technologies. Nuclear power is a source of particular concern because of the possiblility of major reactor accidents; the apparent lack of accepted means for the long-term management of radioactive wastes; and the possible international spread of weapons capabilities as a result of the easier availability of plutonium, particularly if breeder reactors and fuel reprocessing are used. Fossil fuels are also a source of great concern because of the long-term implications for climate of a buildup of carbon dioxide in the atmosphere through the expanding use of coal and other fossil fuels over time, and also because of the potentially damaging impacts of acid rain. Even if the possibilities are slight—as they appear to be—that disasters will actually occur, the effects of worst cases (if they were to materialize) could be devastating. On the other hand, national and international institutional and technical approaches either exist or are potentially available for reducing many of these risks to acceptable levels. It is of the utmost urgency that these approaches be vigorously pursued. What is needed, in addition—because public acceptance may be *the* crucial issue—is a major and continuing effort to keep the public fully informed about what is being done and about prospective outcomes of these actions—both the promising aspects and the continuing problems.

In conclusion, the most fundamental choice that confronts the United States over the long run is between accepting and accommodating to energy scarcity (the present posture) or setting out to achieve energy abundance (a feasible long-run goal). Today's pervasive pessimism about long-run energy supply is not consistent with the facts about natural resources. We are not running out of mineral fuels, not even of raw materials from which liquid and gaseous fuels can be derived. There is no reason to believe that the real costs of obtaining energy in useful forms will continue to spiral upward; it is not unrealistic to contemplate a ceiling on such costs over time, at a multiple far lower (less than a doubling, at most) than those already absorbed in recent years. There is also reason to be hopeful that the costs to the environment and human health and safety of continued growth of energy supplies need not be too great to bear, with the use of appropriate technical and institutional modifications. In fact, new technologies are in many significant ways superior to those they would replace.

The missing ingredients are an adequate appreciation of the real potentials and the political will to take advantage of these potentials. We continue to be immobilized by the adversary aspects of the energy-environment conflict, while the domestic and international needs of the future cry out for technical and institutional solutions that will permit forward movement on both long-term energy supply and environmental protection. We are immobilized also because of the confusion created by the constant drumbeat from those who tell us that we must adjust to an era of limited expectations and must share scarcity with the rest of the world because the energy-resource potentials are no longer available to support the rates of growth that have been experienced in the past.

Domestic energy policies in the United States are important to the world both for the near and the distant future. There is a need for immediate, urgent attention to the problems mentioned at the outset—reducing import dependence and import vulnerability, and developing plans and programs for managing supply emergencies in order to preserve a credible U.S. presence on the international scene through the gravely threatening situation of the coming years. At the same time, we must institute those policies and programs that will ensure energy abundance at tolerable costs over the long term—a supremely important goal for both the United States and the world.

8 Energy in the Developing Countries

Richard H. Sheehan

The World Bank is a major source of financing and technical assistance to developing countries; in our fiscal year 1980, the World Bank, together with its affiliates, the International Development Association (IDA) and the International Finance Corporation (IFC), made lending and investment commitments totaling over $12 billion in the developing world. More than $2.8 billion was lent for energy, of which over $400 million was for oil and gas development—quadrupling the level of the previous year. The bank did not start lending for petroleum development until 1977, but since then it has rapidly accelerated its petroleum program, including a gradual expansion into exploration.

The bank is today by far the largest source of public support for energy development in developing countries—particularly for energy sources other than electric power. However, the urgent need of the oil-importing developing countries to reduce dependence on imported energy has created substantial additional investment requirements subject to the availability of funds. The bank is planning to invest up to $25 billion in energy projects over the next five years, in projects with a total cost of over $90 billion, provided it can find enough money.

In 1980 the developing countries will spend almost U.S. $50 billion on importing oil. If their domestic energy production continued to grow no faster than in recent years (about 7 percent annually), their oil-import bill (in constant 1980 U.S. dollars) would rise by 1990 to $110 billion, a level that would add greatly to the difficulty of financing an already large external deficit.

Although the increase in the real price of oil has put considerable strain on the balance of payments of the oil-importing developing countries, it has also provided them with ample opportunities to tap energy reserves of oil, gas, coal, and hydroelectric and forest resources that were previously regarded as uneconomical or of marginal value. Though maximization of energy production between now and the end of the decade and through a vigorous program of energy conservation, the World Bank estimates that these countries could cut their oil-import bill in 1990 by $25-$30 billion (1980 dollars).

In *real terms* the price of oil is now more than five times what it was in 1972; World Bank projections to 1990 indicate an annual increase of 3 percent a year. Now that energy is no longer cheap, its supply and cost must be given due weight in the plans of economic managers at all levels. These considerations apply not only to forms of energy that are traded internationally, but also to energy that is produced and consumed domestically, and to traditional as well as commercial fuels, because the prices, availability, and consumption levels of all forms of energy are inextricably interrelated. Over the next two decades the increased price of oil is expected to cause significant changes in the sources of supply, and to require a large increase in investment in energy production to reduce dependence on imported oil, and greater efforts to make energy use more efficient.

Following the latest round of oil price increases in 1979, the World Bank commenced a comprehensive review of the world energy situation, with particular emphasis on the likely impact on developing countries. This review concentrated on the perspectives for LDCs in the 1980s, the projected growth in demand and its management, the prospects for expanded energy production, and a possible role for the World Bank. The title of the report, which has recently been published, is "Energy in the Developing Countries." This chapter will focus only on the highlights of the report.

Perspective for the 1980s

Situation of the LDCs

In 1978 twenty-eight developing countries with a total population of about 1.500 million were net exporters of oil. Several of them (notably China, Indonesia, and Nigeria) have large populations that could absorb most of their exportable surplus in the next decade unless significant new discoveries are made or alternative sources developed.

The vast majority (ninety-two) of the developing countries depend in varying degrees on imported oil. Some (notably Chad, Ghana, the Ivory Coast, and Pakistan) have the potential to become self-sufficient within the 1980s. Others, such as Brazil, the Republic of Korea, and Turkey, have a growing demand for oil and could become increasingly dependent on imports unless they develop alternative sources.

Among the ninety-two import-dependent countries are sixty-four countries, including some of the poorest, that depend on imports for more than 75 percent of their commercial supplies. Their energy potential is not well known. Even small discoveries of oil, or relatively small scale development of alternative sources of energy, could substantially reduce their dependence on imports.

Many of these countries are already seriously short of fuelwood and other traditional sources of energy. They include a large majority of the low-income countries, most of which are almost totally dependent on oil imports for their supply of commercial energy. These are the countries that face a double energy crisis.

Energy Policies

Appropriate energy policies differ among developing countries. Among those that import oil, the middle-income countries, especially those that are already semi-industrialized, share many of the energy problems of the developed countries. The switch from traditional to commercial forms of energy has already largely taken place, although there remain areas or pockets in which traditional forms of energy still dominate. These are the countries with rapid rates of economic growth, whose energy requirements oblige them to buy large quantities of oil on the world market. Most of their commercial energy is used in industry, power generation, and transport. These countries have been able to finance their imports through a combination of expanded exports and large foreign borrowings, and some of them have taken appropriate price measures to reduce the growth of demand for oil. They have been able to maintain reasonable, albeit reduced, rates of growth while beginning to make essential adjustments in the structure of their economies. Uncertainties about future markets for their exports, the availability and price of oil, and the extent to which further net foreign borrowings will be prudent emphasize the need for these countries to exploit domestic resources more fully and to formulate policies and programs to maximize the efficiency with which commercial energy is used.

Low-income countries derive half or more of their total energy from wood and agricultural or animal wastes. But many of them depend heavily on petroleum for commercial energy and are short of the resources needed to develop their own energy supplies. They too must be concerned with the energy efficiency of their development since investments made today will determine their energy requirements as modernization accelerates. One of the difficult choices they face is how to stop the rapid depletion of forests and soil fertility without unduly stimulating the use, and therefore the import, of petroleum. Essential elements of the solution are afforestation, small hydroelectric installations, more efficient design of cooking stoves, and more use of coal. In the longer term, local applications of solar, biomass, and other renewable forms of energy hold promise of more abundant energy in rural societies at lower economic and environmental costs.

Management of Energy Demand

All countries can substantially improve the efficiency with which energy is used. This is particularly important for the oil-importing developing countries, since it would save imports. Indeed, with effective demand management and conservation policies, supported by appropriate fiscal, pricing, and regulatory measures, the oil-import bill of these countries in 1990 could be 15-25 percent lower than if present trends continue. The greatest scope for raising efficiency is in the main energy-consuming sectors, such as industry and transport. Improved operating and maintenance procedures, training of staff, and installation of energy-saving devices can yield significant savings comparative quickly.

Investment Requirements

Investment in energy development in the oil-importing developing countries will rise to about 5 percent of GNP, or one-third to one-quarter of total investment. Conservatively, this means an average annual investment of $37 billion (in 1980 U.S. dollars) between 1981 and 1983, and more than $50 billion per year between 1986 and 1990 in energy development. This compares with less than $25 billion in 1980. Almost 75 percent of this will be for power development; but investment in oil and gas is estimated to rise from about $3.5 billion in 1980 to about $10 billion by 1990. Despite these large capital requirements, investment in import substitution in this case makes very good economic sense—in almost all the cases that the bank analyzed, domestic production from primary or secondary energy sources was cheaper than the alternative of importing oil.

But the problem of energy in the developing countries is not only a financial one. Technical assistance in planning for energy development is necessary, as is the strengthening of domestic institutions, the acquisition of technology and skills, and an investment climate that will interest private capital in exploring for and developing still untapped resources.

Prospects for Increased Production of Energy in LDCs

Developing countries consume a small share (12 percent) of the world's commercial energy. However, their economies are growing faster than those of the industrialized countries, and their demand for commercial energy, with the rapid growth of cities, industries, motorized transport, and other energy-intensive developments, is increasing faster than GNP. Up to now much of the increased demand has been met by oil, and most developing countries must import all or a portion of their oil requirements. In 1980 the oil-importing developing countries (OIDCs) are expected to consume some 12.4 million barrels a day of oil equivalent (MBOE), of which about half will be in the form of oil and the remainder in the form of coal, gas, hydro, and nuclear power. However, the production of energy in these OIDCs would amount this year to only 7.8 MBOE, leaving a deficit of 4.6 MBOE. By 1990, under present projections, the consumption figure would increase to 22.8 MBOE, whereas the amount produced would grow only to 15.2 MBOE, resulting in a deficit of 7.6 MBOE.

The striking feature of these projections is that, although their oil production is expected almost to double by 1990 and their energy supplies from other sources will increase, the import expenditures of the oil-importing developing countries in real terms are also expected to double. The OIDCs' consumption of oil will continue to increase, although the annual growth rate is expected to decline from the 7.7 percent experienced in the 1970s to 5.8 percent in the 1980s as the result of a slower growth in GNP and the effect of rising real prices.

Investments in domestic energy production must therefore be expanded as soon as possible if import requirements in the late 1980s and 1990s are to be kept within feasible limits. Energy production is also capital intensive, and in all countries careful choices must be made among potential sources of capital and energy to ensure that available capital, both domestic and external, is deployed so as to maximize economic benefits. Since technologies are changing rapidly in response to the high price of oil, assessments of technical feasibility and economic returns will have to be kept under constant review.

Oil

At present, twenty of the OIDCs (Argentina, Colombia, Romania, India, Chile, Yugoslavia, Bangladesh, Pakistan, Albania, Brazil, Turkey, Afghanistan, Ghana, Barbados, Cuba, Guatemala, Cameroon, Morocco, Phillippines, and Thailand) are producing oil at the rate of 2.0 million barrels per day (mbd), up from 1.5 mbd in 1977. By 1990 they should be able to raise production to 2.9 mbd. Allowing for the depletion of existing proven resources, this would require proving and bringing into production the bulk of their currently estimated remaining ultimate recoverable reserves. Experience in major producing countries suggests that this should be feasible if there is adequate investment in exploration and production facilities. Prospects for additional reserves include second-phase discoveries in older established basins, since more intensive exploration is encouraged by the improved economic viability, at both present and foreseeable prices, of small accumulations, better technology and exploration methods, offshore extensions of coastal basins, and deeper drilling. Increases in reserves could also come from enhanced recovery methods, including the recovery of heavy oil. By 1990 this should enable countries that are now producing to raise their production by a further 400,000 barrels a day, provided an early start is made on enhanced recovery and additional investment can be financed.

Oil has been discovered recently in five OIDCs that are not producers at present (Benin, Chad, Niger, Ivory Coast, and the Sudan). Based on an analysis of these and fifteen other oil-importing developing countries in which exploration is active and where the prospects of discovering oil appear promising, production of 500,000 to 1 million barrels a day by 1990 is considered feasible. A vigorous survey and exploration effort, along with the rapid development of commercial discoveries, in the more promising of the remaining OIDCs should make possible the production of up to 1 million barrels a day by the end of the decade.

Over the past ten years exploration has increased considerably worldwide. However, the amount of geophysical and exploratory drilling activity in OIDCs has not increased, and in some areas there has been a decline. Unless the level of exploratory activity can be raised soon, there is little chance of a substantial improvement in these countries' domestic production during the 1980s, given the time needed to mount an exploratory campaign and develop a discovery to the point of starting commercial production.

Electric Power

Roughly half the world's hydropower potential is in the developing countries, totaling about 1,200 gigawatts (GW), of which only 10 percent has been developed. Given the large increases in oil prices, many hydro sites that were previously uneconomical have become attractive. Developing countries are increasingly taking up hydro surveys and feasibility studies in order to exploit these possibilities; but given the long lead time for such projects, the results during the 1980s are likely to be modest. Nevertheless, about 100 gW of hydro capacity is projected to be added over the next decade in some sixty developing countries. At fuel-oil prices in the range of U.S. $20-$25 per barrel, hydropower costing U.S. $2,500-$3,000 per kilowatt of installed capacity can be competitive with oil-fueled steam units or large diesels.

Although nuclear plants have not achieved the high utilization factor long claimed for them, they now seem to have a significant advantage over oil for baseload units. They do, however, require lead times of about ten years. Nuclear plants raise a variety of safety and environmental issues, including the possibility of loss of coolant; the hazards involved in transporting fuels; and the difficulties of processing, storing, and disposing of radioactive wastes. Countries that will have large enough power systems by

1990 should consider now whether they wish to include nuclear plants in their plans for the 1990s, usually as alternatives to coal-fired plants and in the context of current safety and political issues affecting nuclear energy.

Renewable Energy

Renewable energy resources fall into three broad categories: (1) biomass in its traditional solid forms (wood and agricultural residues); (2) biomass in nontraditional forms (converted into liquid and gaseous fuels); and (3) solar, wind, and minihydro installations.

The demand for *fuelwood*, the most important source of traditional energy for residential uses, including cooking, has grown far faster than supply. Whereas villagers once could usually find enough fuelwood near their homes, many now must search for it half a day's walk or more away, and the urban poor must spend large portions of their incomes on fuel. Many developing countries are therefore facing a second energy crisis of immense magnitude, which affects particularly the rural sectors of their economies. As one dimension of this crisis, the forests of developing countries are being consumed at a rate of 1.3 percent of the total forest area, or 10-15 million hectares, a year. Deforestation is most serious in semiarid and mountainous areas, where it causes serious problems of erosion, siltation, and desertification. As fuelwood supplies are exhausted, animal and crop residues are burned, depriving the soils of valuable nutrients and organic conditioning material. The amount of dung now being burned annually is believed to be equivalent to some 2 million tons of nitrogen and phosphorus. A second dimension is that, if all developing-country households now using traditional fuels were to switch to kerosene, developing countries' demand for oil would rise by 15-20 percent.

Although the fuelwood crisis has already reached serious proportions, technically and economically sound means exist both for reforestation and for improving the efficiency with which wood and other fuels are burned. On the order of 50 million hectares of fuelwood would need to be planted in the developing countries by 2000 to satisfy the projected demand for domestic cooking and heating. This would necessitate a fivefold increase over current planting worldwide. The gap between present and required planting levels is large in all regions, but particularly so in Africa. Here it is estimated that planting would have to be increased as much as fifteenfold to ensure adequate fuelwood supplies. In Asia, which already has serious erosion problems, not only must total planting be increased, but special efforts must also be made to combine increases in planting with measures to control erosion.

The conversion of *biomass* to liquid fuel also holds considerable promise for application and development in developing countries. The production of alcohol, particularly ethanol (ethyl alcohol), from certain types of biomass, is a commercially well established technology. Alcohol in the form of methanol (methyl alcohol) can also be produced from wood. Using alcohol as fuel could help to reduce the consumption of petroleum products for transport. Ethanol is produced by fermentation and distillation of carbohydrate materials such as sugarcane, sugar beet, and molasses, of which only a few plants are so far in commercial use (in Brazil, Kenya, and Mali). It can be used to power vehicles either by itself or in a blend with at least 80 percent gasoline. The greater density, combustion characteristics, and octane-boosting effect of ethanol in blends compensate for its lower energy content, so that within limits ethanol can substitute for an equal volume of gasoline with only minor engine modifications. Methanol-gasoline blends are much more difficult to use as vehicle fuels and do not hold promise for the near future.

Solar, windpower, and *minihydro* technologies are a third source of renewable energy for developing countries. A firm technical basis exists for small hydro and wind-power projects, and these appear to be economically attractive for suitable sites; but there has been little recent experience with them, and much more exploration of sites is needed to assess their potential role. Water heating by flat-plate collectors is the solar technology most ready—technically, economically, and commercially—for widespread application. Some developing countries have begun to manufacture their own solar water heaters, and many others could do so. Flat-plate collectors can be an economical source of hot water for residences and industry; they can also provide heat for drying crops and for certain other agricultural uses.

Demand Management

Demand-management policies can shift consumption from lower- to higher-volume uses, reduce the energy cost of output, and promote a switch from more to less costly sources of supply. At the national level this means setting priorities among the principal users of energy—for example, industrial versus commercial or household activities, public versus private transport, energy-intensive versus nonintensive activities—and ensuring that government policies generally are consistent with these priorities. An essential tool in most countries—both developing and industrialized—for increasing energy efficiency is a pricing policy that ensures that, as far as possible, the price of energy in various uses reflects its real economic cost. In many cases the achievement of economic pricing of energy products requires either the removal of inappropriate government-imposed pricing restrictions or adjustments in government policies.

A government can tax some or all energy products to encourage energy conservation or interfuel substitution. Where political and strategic factors suggest that there may be uncertainties in obtaining energy products from abroad, a government may also allow a premium over economic prices in assessing the viability of developing indigenous energy resources. Governments can also use a variety of nonprice controls—such as import restrictions, quantity rationing, or quotas—for a selective short-term intervention in the market for certain energy and energy-related products while fundamental price adjustments are being made. Finally, it may be necessary to impose on government agencies, which are often insensitive to—or insulated from—market factors, various forms of budgeting or rationing to ensure that the energy consumed in public projects and services also reflects its real cost. Administrative and pricing policies need to be backed by education programs for the general public and for particular kinds of energy consumers, to help overcome political resistance to realistic prices for energy and to inform users of available technologies.

Developing countries could achieve substantial benefits by a broad program of demand management designed to increase the efficiency of energy use. Their energy consumption in 1990, currently projected at 30.6 MBOE a day, could be reduced as much as 15 percent by such a program. Not only could some of the benefits be derived within a short time, but many of the measures involved require little or no investment; an example is the removal of government price regulations, which in many countries (both industrialized and developing) prevent the economic pricing of energy products. The greatest scope for curbing the growth of energy demand is in the industrial sector—which is a major consumer of electric power as well as of mineral fuels—mainly through the planning of industrial development, technical improvements in industrial processing, and retrofitting. Sizable energy savings could also be achieved in transport (particularly of oil), in the electric-power industry and in the household sector. Programs of this kind call for difficult political decisions on the part of governments, as well as for administrative and technical skills that are very scarce in many developing countries. This is an area in which the international community, including both aid agencies and private industry, could be particularly helpful.

World Bank Program for Energy Development

In response to the changed situation in world energy markets, the bank extended its energy operations from the power and coal sectors as mentioned earlier, began to assist member countries with petroleum development. Early in 1979 an accelerated petroleum program was approved by the bank's board that included financing for exploration as well as production. The preparatory work and visits to a number of countries, backed by a comprehensive survey undertaken by consultants, confirmed that many countries possess petroleum resources yet to be discovered. It also revealed the lack of information about resources; the very limited survey and exploratory work that had been done; and the scale of technical, administrative, and financial help that would be needed.

Since the inception of the current energy program, the bank has financed eighteen petroleum projects in sixteen developing countries. World Bank lending for petroleum during fiscal 1979-1981 will amount to U.S. $1.35 billion. Forty-nine developing member countries have been visited by bank energy missions, and in thirty-eight of these one or more petroleum projects have been identified. Of the eighteen projects financed, nine are mainly for predevelopment activities, including technical assistance, geological and geophysical surveys, and exploratory or appraisal drilling. Nine projects are mainly for production. The estimated economic rate of return is high, varying from 30 percent to well over 50 percent.

As part of the accelerated program for fuel minerals, bank missions in the last two years have also reviewed the coal sector in ten developing countries (Afghanistan, Argentina, Brazil, Colombia, Indonesia, Madagascar, Mexico, Philippines, Thailand and Vietnam); this coal-sector work has led to follow-up projects in several of these countries. Lending for coal development in fiscal 1979-1981 will amount to about U.S. $300 million.

World Bank lending for the generation and transmission of electric power increased substantially in fiscal 1980. The average annual lending for power of about $1.0 billion in 1976-1978 increased to $2.4 billion in 1980. The bank's lending for electric-power generation is increasingly emphasizing hydro and coal-fired thermal plants. In fiscal 1980 alone the bank financed 4,847 MW of new hydropower capacity, as much as in the previous four years. In that year there were no oil-fired thermal projects, and gas-fired thermal projects accounted for only 6 percent of the total thermal generating capacity financed; by comparison, oil- and gas-fired thermal projects made up 39 percent on average in fiscal 1976-1978. Several large power projects have been successfully cofinanced by the bank in association with official and private funds, and such projects continue to be attractive to cofinanciers.

The lending program currently planned for energy in fiscal 1981-1985 amounts to about U.S. $13 billion, or 17 percent of the bank's total lending commitments planned for the five-year period. This amount comprises about $4.0 billion for oil and gas; $7.6 billion for electric power; and the balance for coal and lignite, renewables and refineries. Based on a country-by-country review of investment needs and opportunities in the energy sector, it is clear that a substantially larger program would be both feasible and desirable. This so-called desirable program, of U.S. $25 billion is U.S. $12 billion larger than the program currently planned and, without additional resources, cannot be financed by the bank. The desirable program would increase bank lending for electric power to $11.0 billion, would double the lending for oil and gas to $8.0 billion, would significantly increase the lending for coal and renewables; it would also introduce new lending for industrial retrofitting amounting to $1.2 billion. Projects under this larger program would produce (or save) energy equivalent to 2.9 million barrels of oil a day, or 9.5 percent of the developing countries' projected consumption in 1990. The desirable program assumes no change in the bank's policy on cost sharing or in its efforts to attract cofinancing. But to accommodate such a program within the currently approved five-year lending program— energy lending would amount to over 30 percent of the total, compared with the current program's 17 percent—would seriously distort the sectoral balance. The bank would no longer have the appropriate capacity to support the other important development objectives of its members, such as those in agriculture and rural development, health and water supply, education and training, industry and transport, without additional funds.

In view of the urgent need to increase the flow of external capital for energy investment in the developing countries, the bank is studying the possibility of establishing a new energy affiliate. Although many of the elements of such a proposal are not yet defined, the principal objective should be clear: additional resources to finance energy investments beyond the levels now proposed for the bank's fiscal 1981-1985 program. The affiliate's resources must be adequate to sustain a level of operations appropriate to the capital requirements of the sector.

Conclusion

The World Bank's board of directors approved the Energy Report in August 1980, and since then the management of the bank has begun to discuss with its member countries the possibility of establishing the new energy affiliate—probably an International Energy Corporation—that would enable the bank to finance the full $25-billion program.

But the bank's program can only partially contribute to an effort of such vast scope. Much of the financing must come from private capital markets, much of the technology from private firms. The bank's program can help mobilize these financial and technical resources by: providing assistance for energy planning in the developing countries; strengthening institutions in developing countries in the energy sector to provide assessments of resource availabilities and priorities; identifying of investment opportunities for financing by suitable participants; and sharing in the financing of high-priority projects. We believe the bank's expanded energy program will serve the interests of all its members—industrialized and developing, large and small, oil exporters and oil importers alike.

9 The Role of Coal

Leslie Grainger

Introduction

There is a growing realization that the demand for fossil fuels, which currently supply more than 90 percent of the world's energy, will remain strong for a long time within a continuing growth in total energy demand. Coal appears likely to become once again the dominant fossil fuel well within the period during which the world still relies mainly on fossil fuels. Thus the supply and optimal use of coal are crucial factors in successful energy policies.

A number of studies have been made recently on the future role of coal. One that recently received much favorable attention was the World Coal Study (WOCOL), conducted under the direction of Professor Carroll Wilson. The WOCOL findings (which indicate what could and should happen) are illustrated in figures 9-1 through 9-3. Figure 9-4 indicates the bracket of possible growth in import requirements, and figure 9-5 shows where the traded coal may come from. Figure 9-6 indicates the current competitiveness of coal with oil on a thermal basis, even assuming the most stringent environmental requirements. However, as indicated in figure 9-7, the growth in coal demand makes assumptions about the electricity market that may merit further consideration.

The case for coal is based on a strong reserve position and on the flexibility with which it may be used; the flexibility is important because it allows utilization to be adjusted to the needs of the time.

Reserves, Production, and Trade

The most authoritative statement on resources and reserves is that of the World Energy Conference (WEC), illustrated in tables 9-1 and 9-2. These illustrate that proved economic reserves will last a very long time and that the potential for further translation of resources into economic reserves is considerable.

The 1977 WEC survey (before recent pressures) estimated world output in 2000 at 5,780 million tons of coal equivalent (MtCE) compared with the WOCOL high case of 7015. The two estimates are compared in table 9-3. The comparison between 1977 WEC and WOCOL for export potential is

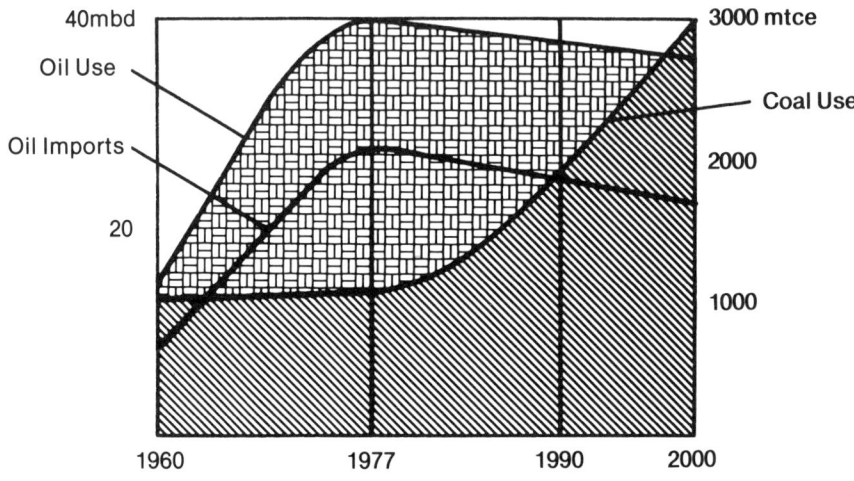

Figure 9-1. Required Coal Expansion, OECD

given in table 9-4. For the year 2020 the WEC predicted an output of 8,846 mtce and exports of 788 mtce. The output would be sustainable for a long time with only a modest translation of resources into economic reserves, presumably by a small shift in prices. The difference between WOCOL and WEC is mainly one of timing; the general feasibility of a three- or fourfold increase has not been doubted for several years.

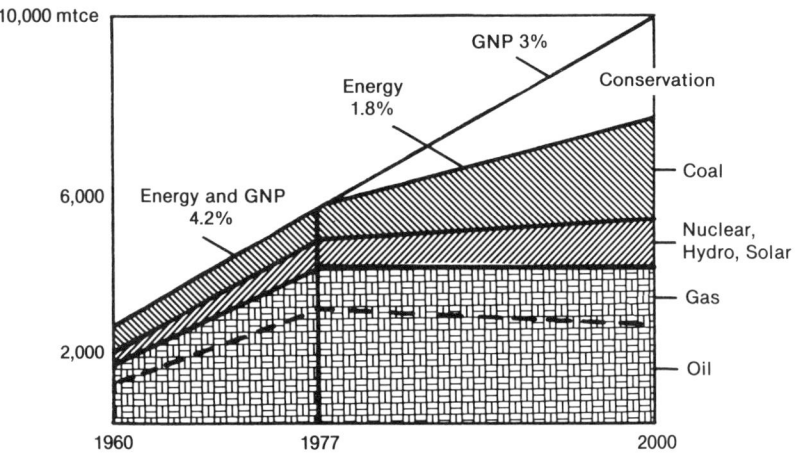

Figure 9-2. Coal's Role in Energy

The Role of Coal

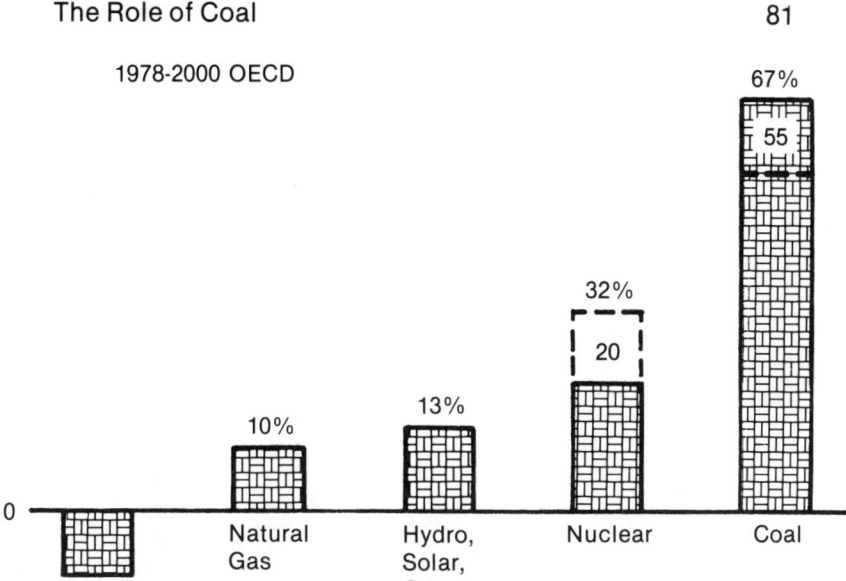

Figure 9-3. Coal's Share of Future Energy Increase

Utilization

Combustion will remain the principal use for coal for some time. Economic comparisons between coal and nuclear power for electricity have been conducted for many years without a sufficiently clear margin being established to justify the elimination of either method for high-load factor plant, especially where environmental controls are not strict. Where full pollution-control procedures must be adopted, the economic penalty for coal plant, by presently applied methods, may be 20 percent or more; however, reliable plant for these purposes can now be purchased. A cheaper alternative may be pressurized fluidized combustion, now being developed; an important application may be "in-filling" a nuclear baseload system with smaller, medium-load factor Central Heating Plan (CHP) plants. Atmospheric-pressure versions are rapidly being applied commercially, and by this means coal is likely to make a substantial and rapid penetration into industrial heating markets.

The technical feasibility of gasification and liquefaction processes is beyond doubt, but large-scale demonstration of second- or third-generation processes is awaited. These newer processes, however, are unlikely to have enormous economic advantages. A factor of about 2.5 times seems a minimum target for the energy cost of coal gases and liquids compared with solid coal. Progress from first-generation processes is likely to be measured

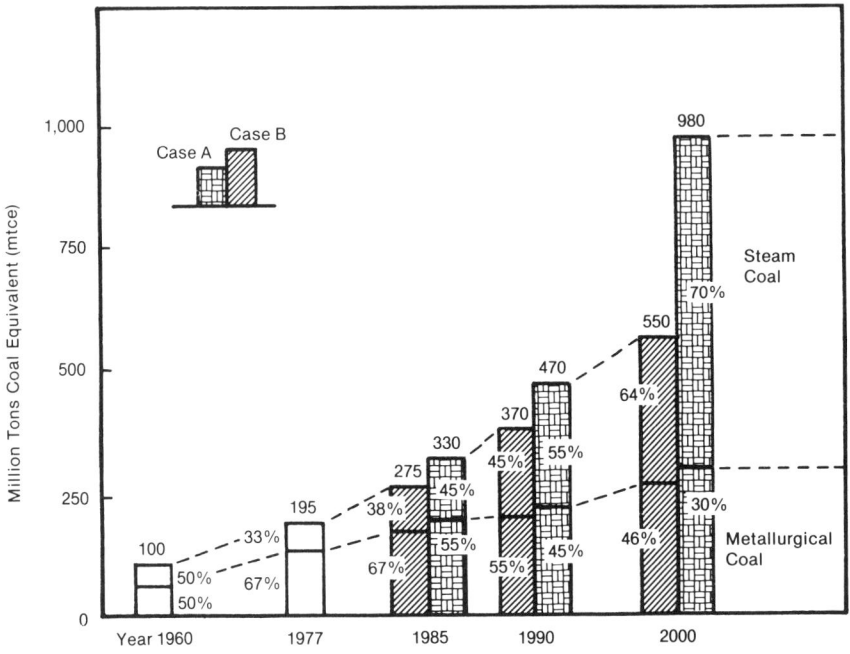

Figure 9-4. World Coal Import Requirements, 1960-2000

by cost reductions of a few percent. Thus the adoption of coal-conversion processes on a major scale seems to depend on a perception of the future relationship between coal and oil prices. To some extent, however, oil prices are likely to reflect the cost of alternatives.

The choice of conversion processes for early commercial adoption is very complex. The achievement of economic comparability may be hastened by the adoption of so-called Coalplex philosophies aimed at minimization of unit capital costs and maximization of energy efficiencies. The conversion of coal into a medium Btu/synthesis gas may be the earliest process to be economic in many circumstances; this process also has the advantage of being compatible with other processes that may be adopted later, leading in an evolutionary manner toward Coalplexes.

Potential Obstacles

Infrastructure

It is accepted that expansion of coal output and trade, as envisaged in the WEC forecast, is well within the technical capacity of the mining industry.

The Role of Coal

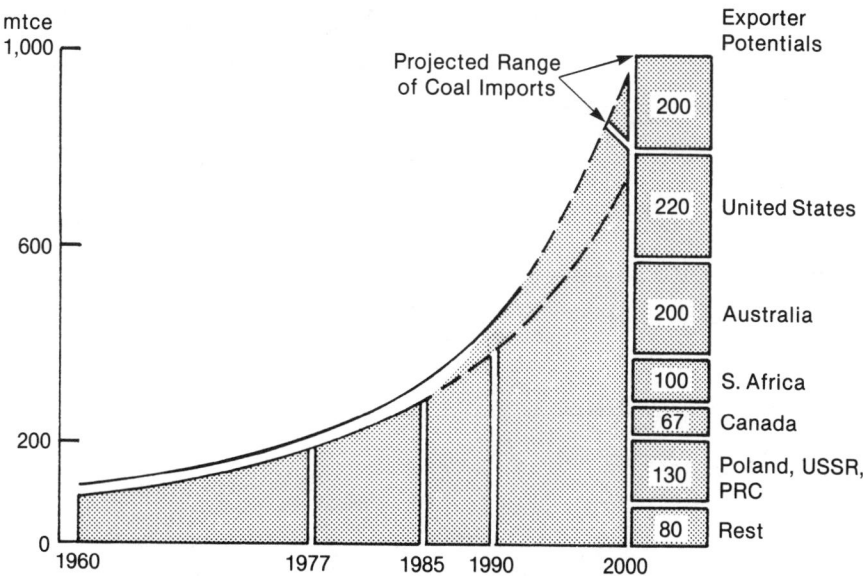

Figure 9-5. Balancing Coal Imports and Exports

The main problems relate to the provision of infrastructure. For example, the WOCOL study estimates the average capital cost of several elements in coal chains as follows:

Mines	$53 per annual tce
Inland transport	$23 per annual tce
Ports	$23 per annual tce
Ships	$59 per annual tce

Of the projected expansion of output, only about 30 percent is expected to enter world trade and hence to require the last two items. The total investment required for the expansion is about $1,000 billion, of which about three quarters will be in user plant, which will be required whatever fuels are used. An important consideration is the timing of investment decisions in infrastructure and user facilities, which require earlier decisions than do mining investments because they involve authorities less directly concerned with the "sharp end" of energy supply.

Environment

There seem to be no environmental problems related to the mining, transport, and use of coal that are technically intransigent. The economic

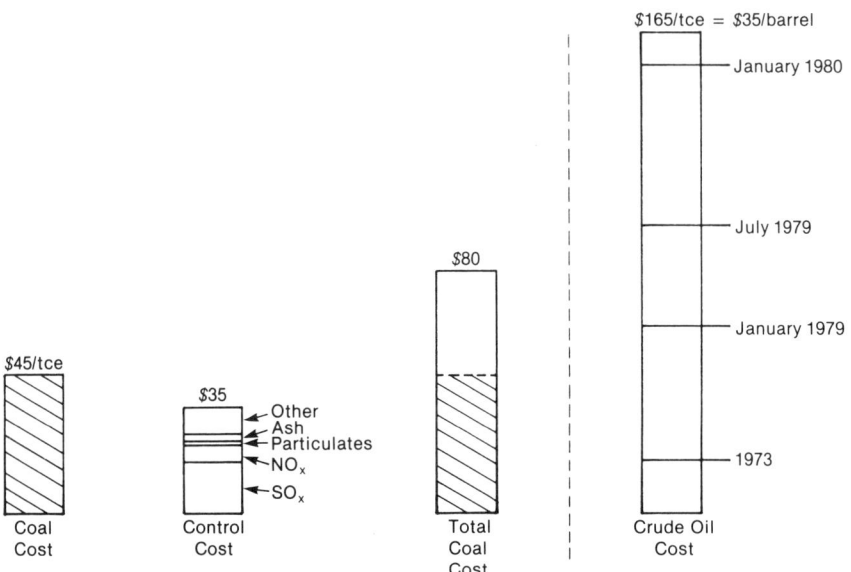

Figure 9-6. Environmental Control Costs, Japan

penalties can, however, be substantial, with the rate probably escalating rapidly as suppression requirements approach the limits of the possible. It is extremely important that environmental standards evolve steadily and at a reasonable rate, consistent with the acquisition of information about harmful effects, quantified in economic terms where possible. An example, quoted at the Nice conference, was the experience with the toxic effects of coke-oven emissions, concerning which earlier fears have not by any means been fully justified by subsequent careful studies. A substantial program of research and development is needed on these problems. A special case is that of CO_2, discussed at some length at Nice.

Conclusions

Recent meetings have confirmed the technical and economic background to the future expansion of coal. The International Conference on Coal Research demonstrated the rapid progress that is being made on all phases of the "second coming" of coal, from mining through utilization. The WEC placed strong emphasis on the need for coal, especially as a complement to nuclear power. The environmental problems were particularly mentioned; a consensus appeared to support the view that environmental problems were important but manageable.

The Role of Coal

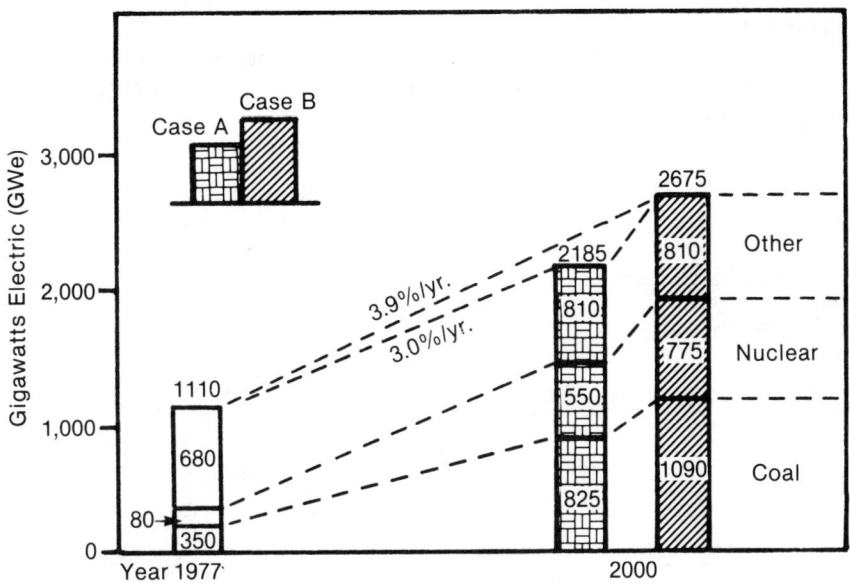

Figure 9-7. OECD Electricity Capacity Mix, 1977-2000

There was also a general consensus that the WOCOL estimates of coal requirement—doubling by 1990 and tripling by 2000—now appeared too optimistic. There was much concern at both meetings over the current supply/demand situation. Having responded to urgent calls for expansion in the wake of the oil crisis, mining companies are finding themselves with ex-

Table 9-1
Development of the Estimates for Geological Resources and Recoverable Reserves of Coal since 1974

Year	Geological Resources in Situ		Technically and Economically Recoverable Reserves		Recoverable Reserves in Percentage of the Geological Resources
	Gt	Gt ce	Gt	Gt ce	
WEC, 1974	10,754	8,603	591	473	5.5
WEC, 1976	11,505	9,045	713	560	6.2
WEC,[a] 1978	—	10,125	—	636	6.3
This report, 1980	13,476	11,062	882	687	6.2[b]

[a]Peters and Schilling, 10th World Energy Conference, Istanbul, 1977 (1978).
[b]For resources and reserves in ce.

Table 9-2
Total Resources of Solid Fossil Fuels in Situ, and Recoverable Reserves (in Gt and Percentage), and the Calculated Lifetime of These Reserves Based on the Production Level in 1978

	Coal			
	Bituminous Coal and Anthracite Gt (%)	Subbituminous Coal Gt (%)	Lignite Gt (%)	Total Gt (%)
Proved reserves	774.6 (58.7)	221.6 (16.8)	323.5 (24.5)	1,319.7 (100.0)
Additional resources	6 161.4 (50.7)	3 835.2 (31.5)	2 159.6 (17.8)	12,156.2 (100.0)
Total	6 936.0 (51.5)	4 056.8 (30.1)	2 483.1 (18.4)	13,475.9 (100.0)
Proved recoverable reserves	487.7 (55.3)	143.0 (16.2)	251.1 (28.5)	881.8 (100.0)
Calculated static lifetime, years	198	780	281	—

cess capacity and unemployed miners. Concluding the discussion at the appropriate session of the WEC, I expressed the view that coal output would continue to be demand constrained. Furthermore, growth in the electricity market for coal would not support the growth in output capacity being proposed, especially if nuclear power were to be expanded at a reasonable rate. Demand constraints would therefore be eased only by the establishment on

Table 9-3
Projected World Coal Output
(m. tonnes c.e.)

	1977 Actual	2000 WEC 1977	2000 WOCOL High	Range of Increase
United States	560	1,340	2,033[a]	780-1,473
USSR	510	1,100	1,100	590
China	373	1,200	1,450	627-1,077
Australia	76	300	366[a]	224-290
Canada	23	115	179[a]	92-156
South Africa	73	233	253[a]	160-180
Poland	167	300	313	133-146
Western Europe	287	410	426	123-139
Others	381	782	895	401-514
Total	2,450	5,780	7,015	3,330-4,565

[a]Including expanded export potential.

Table 9-4
Range of Possible Coal-Export Availabilities
(m. tonnes c.e.)

	1977 Actual	2000 WEC 1977 Report	2000 WOCOL High Case
United States	50	90	350
Australia	38	180	200
Canada	11	40	67
South Africa	12	55	100
	61	275	367
USSR	25	50	50
China	3	30	30
Poland	39	50	50
	67	130	130
Others	12	87	83
World total	190	582	930

a large scale of the new coal-conversion processes. The timetable for demonstration plants followed by the construction of many new plants would not appear to support the WOCOL growth rates.

It is generally accepted that coal growth is likely to be dependent on political will rather than physical or economic difficulties. It is clear, however, that the expression of that physical will relates to long-term, complex issues, which are nonetheless extremely important.

10 The Role of Electricity as a Substitute for Liquid Fuels

L.G. Hauser

Long-range forecasting has become a very sophisticated science. Most industrial executives, when they ask their forecasters for a prediction of the future, want a detailed numerical forecast. A simple forecast—"business will be better in 1990 than it was in 1980"—is of little value to them in their planning. Instead, they need to know how much better, what kind of business will be available, and—in the case of energy—how the energy supplies will divide among the various sectors. Hence the forecaster, in self-defense, starts to build mathematical models in order to supply the requested numbers. These models are becoming more sophisticated each year as more and more interrelationships are built into the model equation. Therein lies one of the major pitfalls involved in using the output from models.

There may be several hundred separate independent assumptions incorporated into a model. These assumptions are seldom transmitted to the ultimate user of the model output. Thus the model output can be misunderstood by the ultimate receiver of these figures because he is not aware of the specific assumptions incorporated in the output.

There is a tendency of forecasters to increase the apparent number of significant figures in their output results. For example, many forecasts of energy demand for 1990 or 2000 are expressed in what appear to be five significant figures. Obviously, no model can be that precise in foretelling future events. In fact, in recent history the mathematical models have not been particularly accurate in forecasting unusual events. For example, no published model predicted the rapid oil price increase in 1973-1974 resulting from the disruption of oil supplies. Nor did any published model, even though the previous history of the 1973-1974 period was available, anticipate the doubling of world oil prices in 1979.

Mathematical models are unsuitable for long-range forecasting because they do not contemplate incipient changes in fundamental relationships. Consequently their outputs seldom predict the significant changes in trends that the real world so often presents.

It has been suggested that a more useful type of long-range forecast should be based on the identification and logical analyses of peripheral events that will influence the trends that are the subject of the forecast. I suggest

89

that this type of forecast be called the *philosophical forecast,* as opposed to the *mathematical-model forecast.* It is my intention to support the thesis of this chapter with philosophical forecasts.

An example of a philosophical forecast is shown in figure 10-1, a forecast of free-world energy demand through the year 2000. It shows some very significant changes in the distribution of energy use among the two segments of the free world. Note that the developing nations of the world go from a 22-percent share of free-world energy to approximately a 60-percent share in the year 2000. This prediction of energy demand is not the result of a mathematical model, but rather contemplates the effect of world population growth, as indicated in figure 10-2. The developed nations are specific ones—namely, the industrialized Western European nations, the United Kingdom, Canada, the United States, and Japan. The balance of the free world is listed under developing nations in accordance with U.N. definitions. The significant factor is the 1.6 billion people who will be the net additions to the population of the developing nations. There will be 1.6 billion people alive in the year 2000 over and above the number present in those countries in 1975. These U.N. population projections are relatively accurate since birthrates and death rates for given societies do not change drastically in a short span of twenty-five years.

The next question is, "What will be the energy requirements of the developing nations with this much larger population?" It is interesting to note that for the period 1950-1975, the growth rate in energy use per person

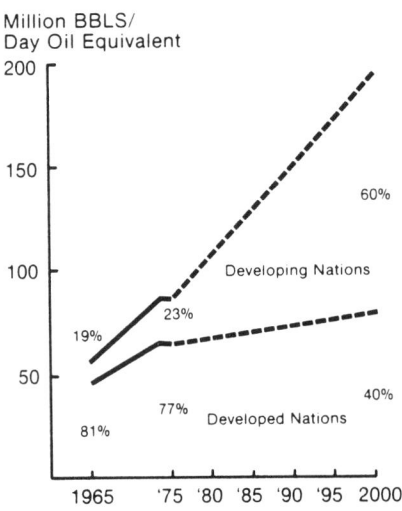

Figure 10-1. A Forecast of Free-World Energy Demand through 2000

Electricity as a Substitute for Liquid Fuels

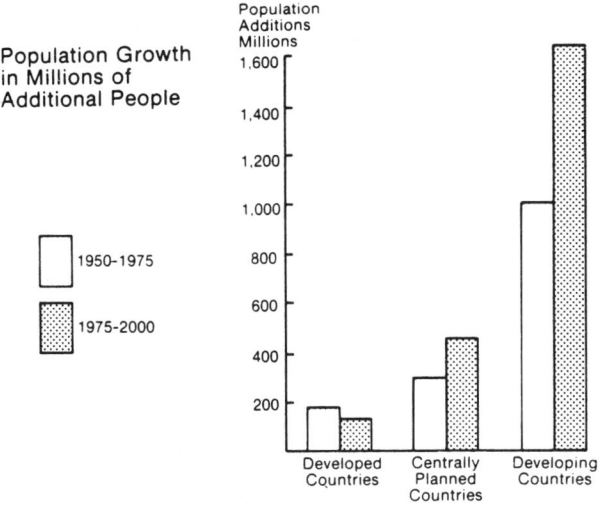

Figure 10-2. A Forecast of World Population Growth to 2000

per year for the developing nations was almost twice that of the developed nations, 4.83 percent compared with 2.86 percent for the developed countries. Hence the question becomes whether the growth rate per capita per year for the developing nations will continue at its historical rate. Several factors indicate that demand may very well continue at its historical pace. The first of these is diet. Figure 10-3 shows the average diet for developed nations as well as for developing nations and shows the specific average diet

Figure 10-3. Average Diets for Developing and Developed Regions

for different geographical sectors of the developing nations. Asia and Africa are the areas of the world in which the diet is poorest. These are also the areas of the world with the largest population growth. In these areas the production of food for simple survival becomes a major problem. Furthermore, as Figure 10-4 shows, the percentage of the labor force devoted to agriculture in the developing nations has reached major proportions compared with figures for the United States or Europe. In the case of Africa, where 70 percent of the labor force is devoted to raising food, it becomes immediately apparent that adding more labor to the agricultural labor force is probably not possible and certainly not desirable. This means that agriculture in the developing nations must undergo some fundamental changes if these countries are to produce the food necessary to feed their growing populations.

Between 1945 and 1970, yield in corn production in the United States increased from 35 bushels per acre to 81 bushels per acre, an increase of 138 percent. Energy used per acre increased by 225 percent. From these figures it is apparent that U.S. agriculture increased and optimized its production per acre per person rather than optimizing production per unit of energy input. Increased production and reduced labor were accomplished by the use of more energy per unit output of food. It is not unreasonable to expect the developing nations, when they develop their agricultural systems, to follow a pattern somewhat similar to the U.S. experience. It will take more energy per unit of food output, which translates into more energy per person, just to maintain the present low level of nutrition. Improvement in diets will obviously require even more energy.

A second major influence on energy demand in developing countries is the population shift to the cities. At present approximately 22 percent live in

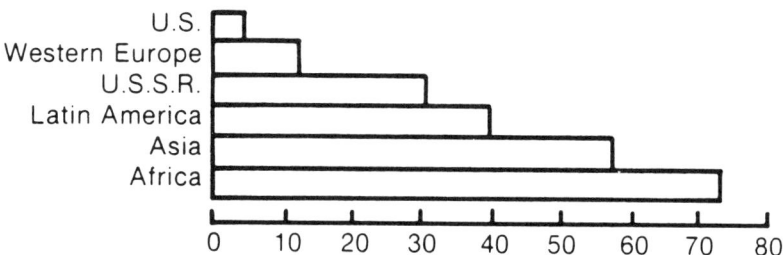

Figure 10-4. Percentage of the Labor Force in Agriculture throughout the World

urban centers. This is expected to increase to 45 percent in urban centers by the year 2000. It requires more energy per person to keep an individual in an urban center than on the primitive farms that now exist in a great many of the developing nations. To supply the bare necessities of city life requires energy for water, sewage, electricity, heat, and transportation. This must be compared with the needs of the primitive farm family living off the land and requiring essentially no outside energy supplies. Thus at least two fundamental shifts in the developing nations are going to require more energy per person—the production of food and the shift to the cities.

The forecast in figure 10-1 is based on the assumption that the annual growth in energy consumption per person per year in the developed nations would be zero, and that the energy growth in the developing nations would continue at its historical rate of 4.86 percent per year per person because of the factors just described. The important output of this forecast is not the numerical results generally associated with the forecasts, but rather two fundamental conclusions: first, the demand for energy by the developing nations will continue to grow at significant rates; second, by the year 2000 the developing nations will be the largest users of energy in the free world. This second point will have material impacts on world trading partners and on the international balance of power. However, discussion of these aspects is outside the scope of this chapter.

Many objections have been offered to this forecast on the grounds that the developing nations will not have the wealth available to purchase this major increase in energy supply. That may be, but there is another factor that should be considered (see figure 10-5). The average annual use per person of energy in the developed nations is 35 barrels per year of oil equivalent. At present, the average annual use in the developing nations is only 3.6 barrels per year per person, or roughly one-tenth that of the developed nations. It is likely that their growth rate will continue through the end of the twentieth century, when they will be using 11.6 barrels per person, or one-third that of the developed nations. The argument that the developing nations will not have the wealth to purchase the required energy has prima facie merit. However, countervailing forces must be considered. The fundamental need for energy by the developing countries is for life survival. In effect, they are in a position similar to that of the disadvantaged minorities in the developed countries two or three decades ago. There has been a definite shift in policy in the distribution of wealth in the developed nations in the recent past. It is not unreasonable to expect this distribution of wealth to continue in other nations of the world in the foreseeable future. In fact, a recent announcement by Mexico and Venezuela supports this trend; they announced that they would make suitable financial ar-

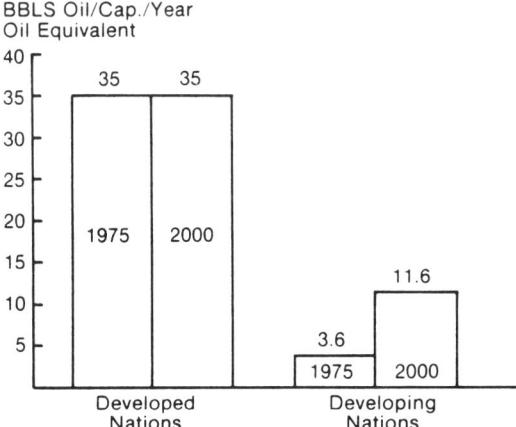

Figure 10-5. A Forecast of Free-World Energy Demand in Barrels of Oil Equivalent per Capita per Year

rangements with the "have-not" nations in Latin America to guarantee them petroleum for their essential needs. Thus one must conclude that although strict economic analysis would deny the growth in energy use to the developing nations, social pressures and general historical trends on the distribution of wealth would indicate that the developing nations' share of energy per person will increase relative to the present share in the developed nations. These fundamental trends are normally not addressed in typical mathematical models of energy demand.

Another major point to consider is that energy demand by many of the developing nations can be satisfied only by an energy source that is easy to use, easy to store, and easy to transport. Oil is the only fuel that meets all these requirements. However, it is well recognized that the hydrocarbon resources of the world in the form of oil and natural gas are likely to be in short supply compared with demand before the turn of the century. This makes the problem even more serious because the developing nations that need this energy for their very survival will need a major portion in the form of oil. Thus it becomes imperative that the developed nations consider what other sources of energy they can use to offset their current demand for oil.

This chapter will address some of the fundamental changes that will occur in U.S. energy demand that could have a major influence on the demand for oil in this country. One of the major ways to examine the future demand for energy is to identify the fuel options that the ultimate user has available today. These are as follows: oil, natural gas, coal, electricity, and

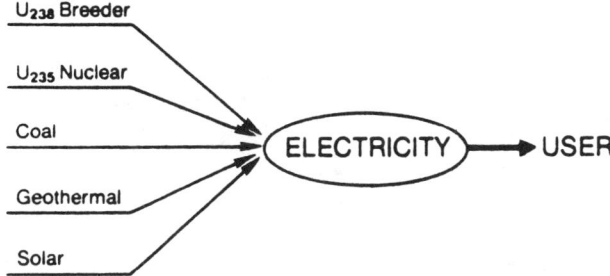

Figure 10-6. Utilization of Potential U.S. Energy Resources by Users of Electricity

solar. The first two, oil and natural gas, are options today; but it is a well-accepted fact that future supply of these two fuels will not be adequate to meet total world demand. Hence a user does not really have an option in the long run of using either of these two fuels. That leaves only three: coal, electricity, and solar.

Some may find fault with the classification of electricity and solar as fuels. But from the user's viewpoint, these three options supply him with energy; he would consider them the "fuels" he must pay for in order to obtain the energy he desires. There are many advantages to the user in selecting electricity. Often the major consideration of energy planners is the capital investment in the energy-producing equipment. But from the viewpoint of society, the main factor may be the capital investment in the energy-utilizing equipment. For example, if a person is currently using an oil furnace to heat his home, he has capital invested in that furnace. If oil becomes unavailable to him, it is necessary for him to select a new type of furnace, say a gas furnace, and invest capital in the gas furnace. If within relatively few years gas is no longer available, it is then necessary to invest in a third furnace, such as a coal furnace, in order to utilize that source of energy. As shown in Figure 10-6, if the user invests in electrical equipment, such as a heat pump, then he need not reinvest capital in his energy-utilizing equipment if the prime source of energy is changed. If the electric utility, for example, changes from producing electricity from coal to producing electricity from uranium, it has no impact on the user's capital investment. This fundamental advantage of electricity as a fuel is usually not considered in most economic evaluations of fuel options.

To be more specific in addressing the substitution of electricity for liquid fuels, there are four distinct areas that deserve consideration.

Electric Heating

The United States currently uses approximately 40 percent of its oil to produce heat in stationary boiler applications. The oil is burned for its heating value only in household heating, industrial-process heat, power generation, and manufacturing operations. There exists today electrical apparatus that can perform these same functions using electricity as the fuel instead of oil. For example, in household heating the heat pump is an ideal application to replace oil furnaces. In manufacturing there are resistance boilers and ovens, induction furnaces, and arc furnaces to produce manufacturing heat and process steam using electricity as the source of energy. A new product is the self-stabilized arc heater produced by Westinghouse. This device generates heat by establishing an electric arc in a confined controlled space, which then heats the gas that is passed across the electrode faces. This gas can be air, nitrogen, or another process gas for a specific application. Installations have been made in steel manufacturing, and applications in the acetylene industry are under investigation. This provides a high-temperature, clean gas flame with very fine controls on the temperature of the flame. the flame.

Conservation of Energy

Conservation of energy generally is defined as making more efficient use of energy in a process or work application. For example, many of the process industries are increasing process efficiency through more reclamation of previously wasted heat. This involves the use of pumps and fans, which are generally driven by electric motors. It is not at all uncommon for a process to utilize 200 BTUs of electrical energy driving a fan motor to save 1,000 BTUs of primary energy in the process. This in effect is the substitution of the 200 BTUs of electrical energy for the 1,000 BTUs of gas or oil energy. Another way to make processes more efficient is through the use of better controls and sensors, which usually are driven by electricity. New technologies that incorporate miniaturization and hence use less energy also depend on electricity for their effectiveness. Examples such as the microprocessor chip abound in current technology.

Automated Factories

As U.S. industry attemps to reverse its productivity decline, it will be necessary to use more and more robotics and computer controls in the manufacturing process. The fundamental energy source for both these

Electricity as a Substitute for Liquid Fuels 97

technology changes is electricity. A third factor is rapidly assuming prominence in the industrial world—the paperless office. The move toward word-processing centers, magnetic data storage, and computer drafting, involves the replacement of energy-intensive paper with miniwatts of electricity to drive the cathode-ray tubes. These factors are in the forefront of developing technology as we attempt to improve the efficiency of our industry, and these factors depend on electricity in order to perform their functions.

Communications

Rapid progress has been and is being made in the area of personal communications. We have already established in the United States the beginning of an electronic postal service, wherein messages are transmitted by wire in a semiautomatic fashion to a distant point without the delay and use of energy needed to transmit that same message in a written form on a piece of paper. Rapid progress is expected to be made as more and more of the electronic transmission methods are utilized not only in industry but also for individuals in their communications network. In addition, we can look forward in the not-too-distant future to the arrival of the electronic meeting, wherein individuals will not be required to move physically to a common meeting place but instead will be present in the form of electronic televised images, which will be made into a montage representing all members present at the meeting. This obviously involves a major saving in energy because it will not be necessary to spend the fuel to make the physical trip. It will also represent more efficient use of an individual's time. Here again, miniwatts of electricity are substituted for substantial quantities of petroleum fuel in the transportation sector.

The foregoing four areas are significant because they are where electricity will in effect substitute for other forms of fuel. Figure 10-7 is a schematic diagram showing electricity as the cornerstone of the technological progress currently underway in the United States. The question that remains is, "How does the long-range forecaster incorporate these fundamental changes into his forecasting model so that the end result will reflect the impact of these changes?" Often it is helpful to examine recent trends in energy applications. Figure 10-8, for example, shows the abrupt change in the incremental domestic energy supply in the United States, comparing the six-year period prior to the 1973 oil embargo with the six-year period immediately following the embargo. This figure shows the sources of the incremental energy-production increase in the United States for these two periods. The emphasis immediately changed from natural gas, which supplied 60 percent of the incremental energy production prior to the embargo, to coal, which supplied 60 percent of the incremental energy production after the embargo. Nuclear

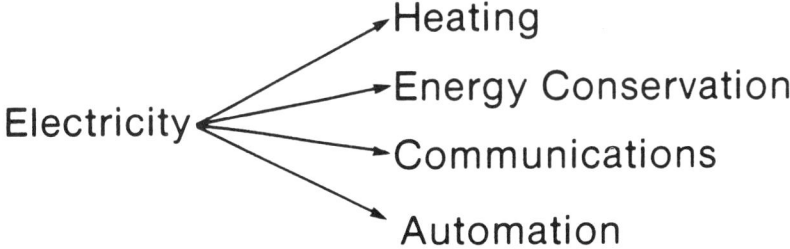

Figure 10-7. Electricity as the Cornerstone of Technological Progress in the United States

assumed a larger role, changing from an approximately 11-percent increase in the earlier period to a more than 30-percent increase in the period after the oil embargo.

The next interesting statistic to examine is the change in energy use by end-use sector in the six years after the 1973 oil embargo, as shown in table 10-1. From this table, it readily becomes apparent that the switch to electric substitution started immediately after the 1973 oil embargo. The only

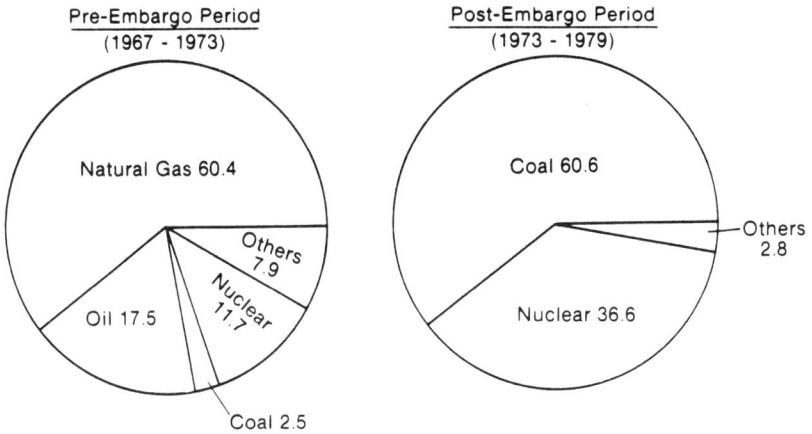

Figure 10-8. Incremental Energy Supply in the United States, Preceding and Following 1973

Electricity as a Substitute for Liquid Fuels

Table 10-1
Changes in Energy Use by End-Use Sector, United States, 1973-1979

End-Use Sector	Electricity (%)	Nonelectric Energy (%)
Residential and commercial	+19.0	−3.5
Industry	+21.6	−7.6
Transportation	—	+8.6
Total economy	+20.2	−0.4

nonelectric energy increase in that resulting six years was in the transportation sector. That might be related to the fact that the prices of liquid fuels for transportation were price controlled during that period. The fundamental reason for this shift might be explained by figure 10-9. In this figure, the real-price increase (without the inflation factor) has been calculated for the same time period—1973-1979. From this it is evident that electricity had by far the smallest real-price increase of all of the fuel options available to the users. It could be argued that this price sensitivity working in the essentially free market of fuel options is primarily responsible for the rapid increase in electrical usage, as opposed to a decline in usage of nonelectric energy.

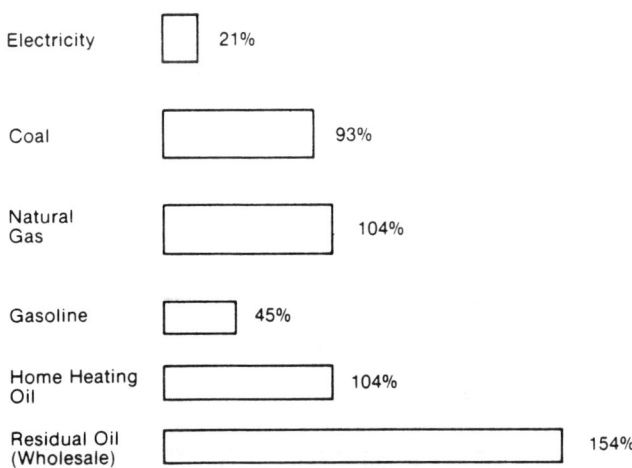

Figure 10-9. Real Price Increase in Energy, 1973-1979

Figure 10-10 shows a forecast of real-price increases for the next twenty years by fuel option. In each case it is apparent that electricity will have a smaller real-price increase than the other options available to users. This leads to the conclusion that the growth in electricity use will continue to be greater than the growth in nonelectrical energy.

This raises the question, "What will be the likely growth rate for electricity?" Consideration of the factors raised in this chapter can give guidance for this prediction.

1. The developing nations will continue to have a growing demand for oil energy to meet their survival needs.
2. Because the normal growth in world oil demand will exceed production by 1990, industrialized nations must reduce their oil consumption to prevent disastrous world unrest.
3. Coal and electricity are the only practical options to oil energy available to the energy users for the required alternative energy supply.
4. The emerging technologies for improved energy-utilization efficiency and for improved productivity depend on electricity.
5. Recent trends in energy indicate that users are choosing the electricity option rather than the other fuels.

A forecast based on the foregoing considerations, figure 10-11, shows the total kilowatt output for the United States by five-year periods through 2000 with an average growth rate approximating 3 percent per year. Figure 10-12 shows the energy consumption by sector for 1980, 1990, and 2000. From this it is obvious that nonelectric energy growth is essentially zero and

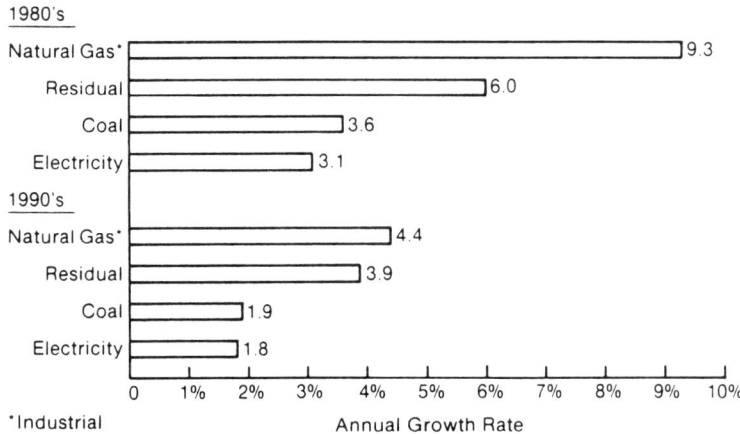

Figure 10-10. Forecast of Real Price Increases in Energy through 2000

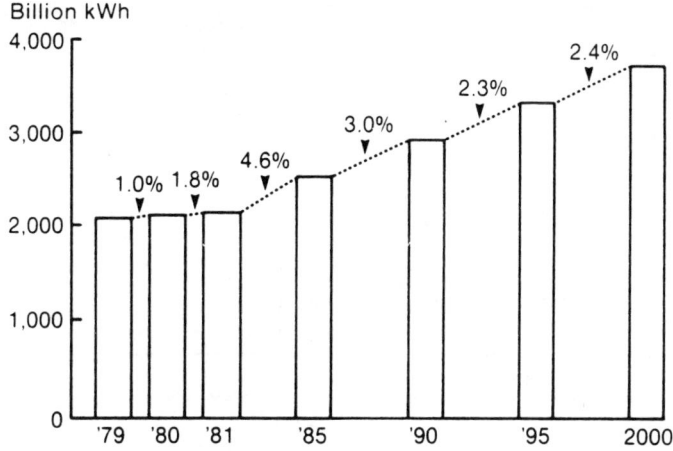

Figure 10-11. Forecast of Electricity Output in the United States through 2000, in Kilowatt-Hours

that only the electrical energy part of the total energy picture is increasing. This is summarized in figure 10-13, which shows that the total growth between 1980 and 2000 for nonelectrical energy is 13 percent, whereas the total growth in electrical energy is 79 percent. Thus electricity will grow at a rate six times that of nonelectric energy.

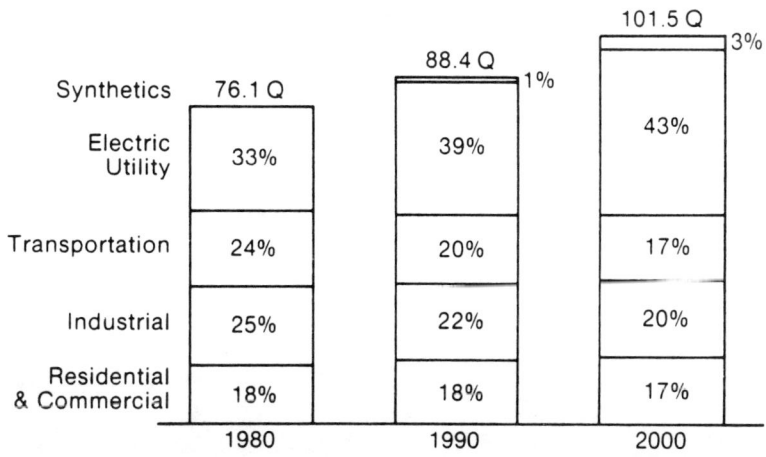

Figure 10-12. Forecast of Energy Consumption by Sector in the United States through 2000, in Quads

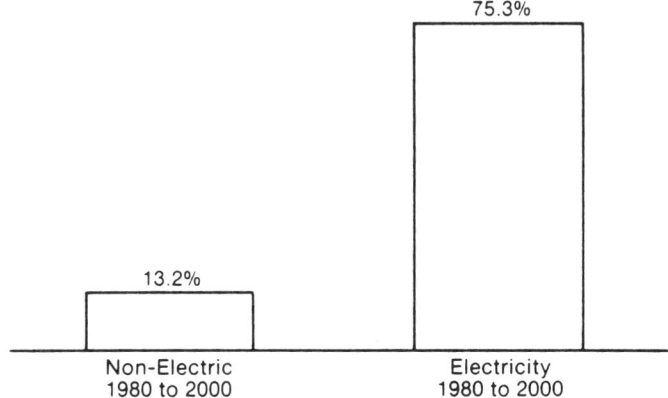

Figure 10-13. Forecast of Increases in Nonelectric and Electric Energy Consumption in the United States to 2000

This philosophical forecast has produced results that are sharply different from several published forecasts based on energy-conversion efficiencies or on short-run economic evaluations. I suggest that the peripheral relationships discussed herein will have a greater impact on future energy trends than will these other factors.

11 The Disequilibrium Effects of Oil-Price Shocks

Narasimhan P. Kannan and *Martin M. Scholl*

Introduction

A major concern of energy planners in the countries that belong to the Organization for Economic Cooperation and Development (OECD) is to insulate their economies from the potentially volatile world oil markets. This concern arises from the fact that recent events, such as the war between Iran and Iraq, have increased the likelihood of interruptions in oil supply from the Middle East. Furthermore, the concern is based not so much on the monopoly power in setting prices of the Organization of Petroleum Exporting Countries (OPEC), which in its recent attempts has failed to achieve consensus among its members, but on the ability of some OPEC members to implement a targeted embargo against the developed countries. One of the vulnerable nations among the OECD members, for various geopolitical reasons, is the United States. Thus this chapter will primarily address the potential impacts of a targeted oil embargo by the Arab nations in the Middle East that are members of the OPEC. Although this analysis is restricted to the United States, the basic principles developed here are equally applicable to other OECD nations.

Any meaningful analysis of the impacts of an Arab oil embargo against the United States must take into account three important factors. First, the domestic price of oil will be completely deregulated by the end of 1981, at which point it is expected to reach the world oil price relatively quickly. Thus the deregulation would expose U.S. oil consumers to the vagaries of world oil prices. Second, the U.S. government currently has 91 million barrels of crude oil stocks in the strategic petroleum reserves, which barely covers two weeks of domestic needs at the current consumption rate of 7 million barrels per day. Finally, we must take account of the degree of U.S. dependence on imported oil in general, and Arab oil in particular.

Table 11-1 shows U.S. oil imports from Arab sources, from OPEC, and from all sources for the last seven years. It suggests three important trends. First, total oil imports by the United States as a fraction of total domestic use grew from 36.2 percent in 1973 to 47.8 percent in 1977 and have declined progressively since then to a current (1980) level of 40.6 percent. Second, oil imports from OPEC countries as a fraction of total oil imports rose

Table 11-1
U.S. Dependence on Imported Oil

	From Arab OPEC Countries		MMBBL/Day of Import From All OPEC Countries		Total Imports from All Countries		Total Domestic Use in MMBBLS per Day
Year	Quantity	Percentage of Imports	Quantity	Percentage of Imports	Quantity	Percentage of Total Domestic Use	
1973	0.91	14.5	2.99	48.0	6.26	36.2	17.31
1974	0.75	12.3	3.28	53.7	6.11	36.9	16.55
1975	1.38	22.8	3.60	59.4	6.06	37.1	16.32
1976	2.42	33.1	5.07	69.4	7.31	41.9	17.46
1977	3.19	36.2	6.19	70.3	8.81	47.8	18.43
1978	2.96	35.4	5.75	68.8	8.36	44.3	18.85
1979	3.04	36.1	5.61	66.7	8.41	45.6	18.43
1980 second quarter	2.57	38.2	4.25	63.2	6.72	40.6	16.54

Source: *Monthly Energy Review*, September 1980, U.S. Department of Energy/Energy Information Administration.

from 48 percent in 1973 to a staggering 70.3 percent in 1977 and declined thereafter to a current level of 63.2 percent. Despite the peaking of these two fractions in 1977, however, the fraction of U.S. imports from Arab OPEC countries has risen steadily from 14.5 percent in 1973 to 38.2 percent in mid-1980. These figures imply that, although U.S. dependence on imported oil has declined over the last three years, the nation has become progressively more dependent on Arab OPEC countries for oil.

These trends raise many policy issues, the most important of which is the impact on U.S. economic growth and employment of a potential oil embargo by the Arab/OPEC countries. To give a dimension of reality to this issue, we can translate this concern into the following specific question: What would be the impact on the U.S. economy of a total cutoff of oil supplies from Arab OPEC countries for a whole year in 1981, or 1983, or 1985, or 1990? Before answering this question, it is useful to review some of the typical dynamic responses of the U.S. economy to a potential shortage of oil.

Major Dynamic Responses to Energy-Price Shocks

Deregulated domestic oil prices mean that any reduction in supply would quickly cause a rise in the prices of oil, given the short-run inelasticity of demand (estimates range between -0.05 and -0.1). Thus a sudden cutoff of oil imports would subject the domestic economy to a price shock. The sudden rise in prices, however, would set into motion the market mechanisms of conservation and increased domestic supply, but these do not occur instantaneously.

It takes anywhere from six months to five years for the effects of oil conservation—through thermal insulation; industrial retrofitting; deployment of energy-efficient capital plant and equipment; use of fuel-efficient automobiles; and substitution of coal, natural gas, and electricity—to be fully realized. The supply of domestic fuels would take even longer, given the time lag included in increasing domestic supplies from coal production, coal conversion, electricity production, and renewable resources.

Thus, if the forces of demand and supply of substitutes for imported oil do not respond fast enough to a sudden and unanticipated oil-price shock, there is bound to be a large-scale transfer of income from the energy-using sectors of the economy to the energy-producing sector, particularly to the oil industry. This would cause a temporary reduction in profitability of nonenergy capital and thus would affect future economic growth. Most important, however, a sudden decline in the availability of oil is bound to cause shortages and plant shutdowns and lead to severe unemployment. All these responses, along with some of the long-term dynamic responses, are summarized in table 11-2.

Table 11-2
Alternative Dynamic Responses to Rising Energy Prices

Sector	Type of Response	Range of Typical Lag Times (Years)	Uncertainties
Energy-production sector (supply sector), partially controllable by government	Increased capacity utilization	0.05-0.1	Very small
	Increased investment in conventional production capacity and new sources	5-10	Depletable-resource estimates
	Increased investment in R&D	10-20	Unpredictability of result
Non-energy-production sector (demand sector) Partially controllable by government	Improved energy efficiency	0.5-5	Uncertainties in costs of aggregate response
	Life-style changes	5-20	Difficult to predict the magnitudes of change
Uncontrollable by government	Reduced investment due to capital profit squeeze and leading to reduced output in nonenergy industries	0.5-5	Short-run substitution elasticities for capital, labor, and energy
	Increased productivity of capital and labor	5-20	Difficult to predict

Table 11-2 shows approximate ranges of time lags of each of these short- and long-term responses. Some of these can be controlled by government through the conventional policy levers of tax incentives and regulation. However, there is little government can do to respond quickly to sudden shortages imposed from outside sources except to allocate by rationing.

In order to test the implications of all the responses listed in table 11-2, one needs a computer model of the U.S. energy/economy system, as the complex interactions become intractable otherwise. One such model, known as the ENERGY1, was used in testing for implications of an embargo.

The ENERGY1 Model

The ENERGY1 model is specially designed to analyze the impact of alternative national energy policies and contingencies. It consists of two separate models: ECONOMY1, which represents the nonenergy sector of the U.S.

Disequilibrium Effects of Oil-Price Shocks

economy, and FOSSIL79, which represents the domestic energy-production sector. The basic interactions between these two models are shown in figure 11-1.

Given a set of exogenous specifications such as world oil prices, U.S. population and labor-force participation rates, supply curves for energy sources, and costs of energy-efficiency improvements, the model can be used to test the dynamic implications of various energy policies on GNP, nonenergy output, and outputs and prices by source of energy. This allows an analyst to determine the likely trade-offs between economic growth and energy self-sufficiency.[1]

Some of the major assumptions embedded in the structure of the ENERGY1 model are as follows:

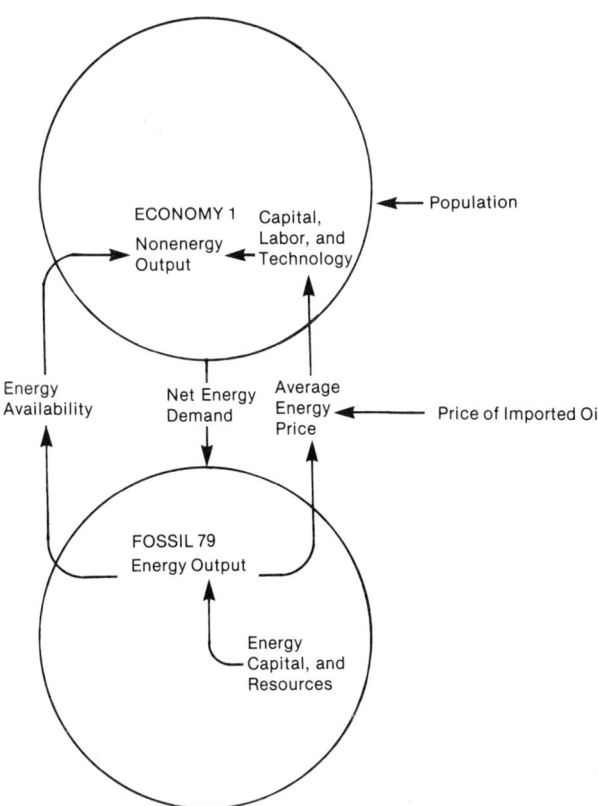

Figure 11-1. A Schematic of the Basic Interactions in the ENERGY 1 Model

1. The U.S. government follows full-employment policies.
2. The values of exports and imports are equal.
3. The demand for oil imports is a residual of domestic demand and domestic supply.

Analysis with ENERGY1

The ENERGY1 model was calibrated to incorporate the most recent energy policies enacted by the Congress. This provided a base case for comparing subsequent model runs with the imposition of an Arab oil embargo.

The base-case population, labor-force participation, and the world oil-price assumptions are shown in table 11-3. In addition, it was assumed that domestic oil prices would be completely deregulated by September 1981 and that the provisions of President Carter's synthetic fuels bill would be implemented.

Equipped with these assumptions, the ENERGY1 model was run to project the future response of two criteria variables—nonenergy output and domestic average energy price—under the following five scenarios:

1. base case;
2. a 40-percent cutoff in imported oil (to reflect an Arab/OPEC embargo) lasting for one year;
3. a similar embargo to the previous case in 1983;
4. an embargo in 1985;
5. an embargo in 1990.

Table 11-3
The Base-Case Assumptions

Year	Population: 10^6 Persons[a]	Employed Labor Force[b]	World Oil Price (1975 $/BBL)[c]
1980	222.0	102.80	23.00
1985	233.5	112.10	26.10
1990	245.0	120.05	31.90
1995	253.0	126.50	37.12
2000	262.0	131.0	40.6

[a]Bureau of Census Series II projections.
[b]Estimates based on a linear growth in labor-force participation rate from the current 48 percent of population to 50 percent by 2000, and a normal unemployment rate of 6 percent through the next twenty years.
[c]These values were assumed to represent an optimistic case of world oil-price behavior.

The base-case projection of the two criteria variables is shown in figure 11-2. In this case the nonenergy output continues to grow at 1.6 percent per year between 1980 and 2000, reaching a level of $2.25 trillion (1975 U.S. dollars) by 2000. The domestic average energy delivered price (a weighted average of all energy prices by fractional use) grows sharply during the early 1980s, following deregulation, and then gradually tapers off at a level of $7.4 per MMBtu (1975 dollars) by 2000. The letter E in the plot in figure 11-2 indicates primary energy consumption, which reaches a rate of 92 quads per year by 1990 and stays at that level until the year 2000. Although it is not shown in the figure, oil imports are projected to decline to a level of 11 quads by 1990 and to 4 quads by 2000.

In figure 11-3, the projections from ENERGY1 model runs are shown for the four Arab/OPEC embargo cases. The most striking feature of these projections is that an embargo causes a sudden and sharp rise in domestic

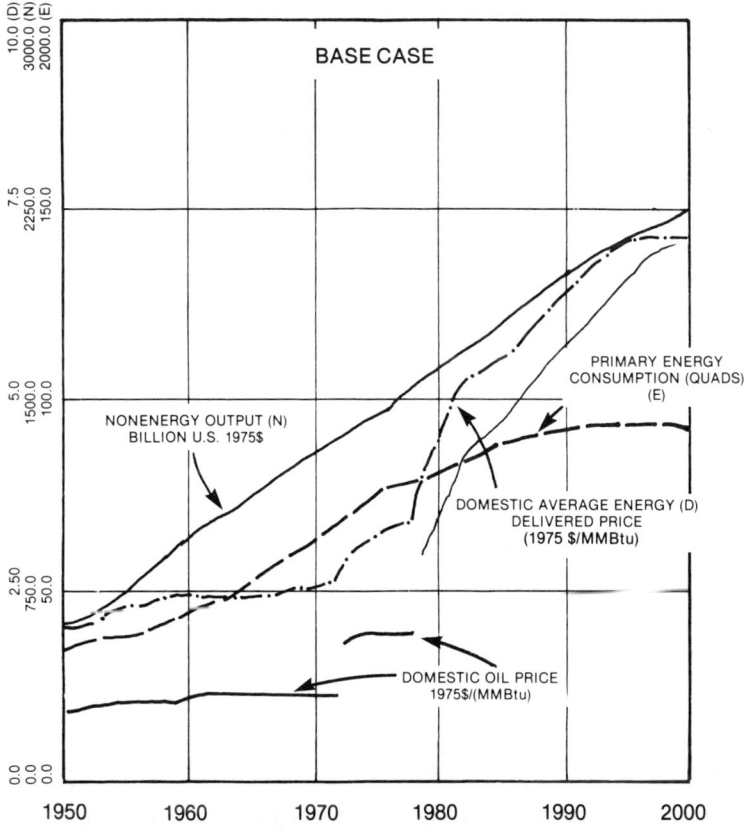

Figure 11-2. The Base-Case Projections from ENERGY 1

average energy price, which in turn causes a correspondingly sharp drop in nonenergy output. Furthermore, the projections suggest that an embargo, if it occurred in 1990, would cause a more severe drop in output than if it did so in 1981, 1983, or 1985.

This last inference is crucial. In order to explore this further, consider the projected drop in outputs shown in table 11-4. This loss of nonenergy output increases with the year in which an embargo is imposed. This is partly due to the fact that the short-run elasticity of substitution for imported oil declines with time as the economy exploits the most cost-effective efficiency improvements in the initial years. For instance, as we get lower on the demand curve for energy, the price elasticity declines—that is, the marginal

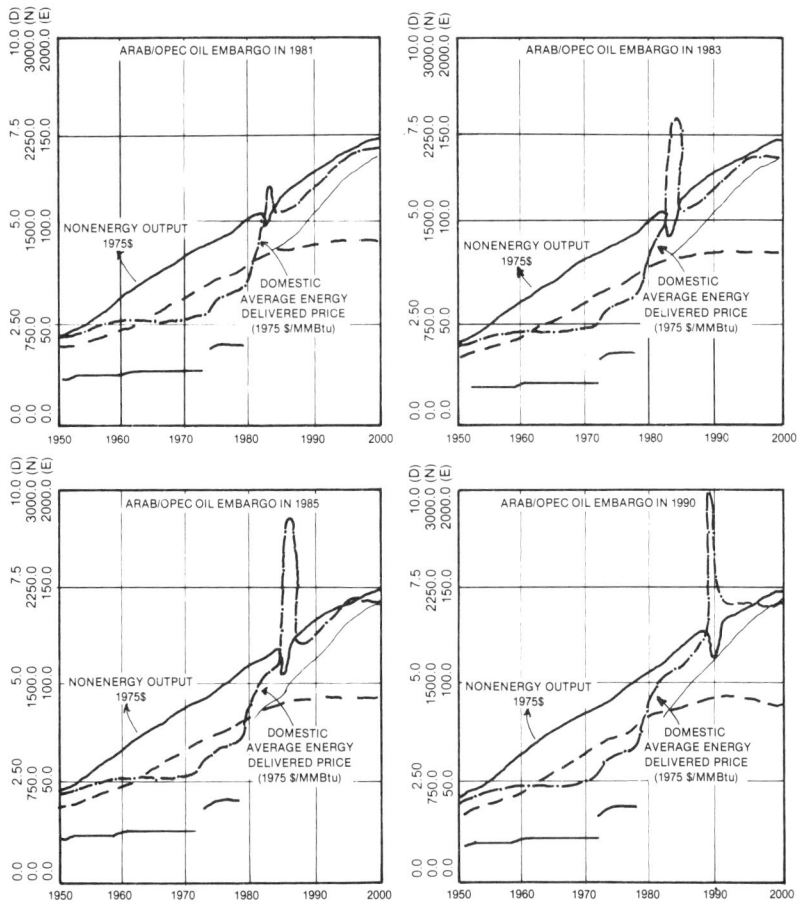

Figure 11-3. ENERGY 1 Projections for Four Cases: Embargo in 1981, 1983, 1985, and 1990, Respectively

cost of efficiency improvement will increase with time. This is perhaps the reason that the energy-efficient economies of Western Europe would be more vulnerable to an oil embargo than would the U.S. economy.

Beyond the demand-side constraints, the new energy supplies from synthetic fuels and renewable resources will not contribute significantly to U.S. energy needs before 1990. Thus the U.S. economy would become progressively more vulnerable to an embargo in the 1990s even though the magnitude of domestic imports is likely to decline. This is perhaps the most significant conclusion to emerge from this analysis.

In order to give an added dimension of reality to the magnitude of impacts, the estimates of increase in unemployment rates are shown in table 11-4 for each of the four cases. An increase of 15 percent in unemployment rate in 1981 would imply loss of work for over 20 million individuals in the domestic labor force. And a 23-percent increase in 1990 would mean a loss of 30 million jobs, a case of severe depression.

Summary and Conclusions

A successful oil embargo implemented by the Arab/OPEC countries against the United States would have severe consequences; the later the embargo were imposed during the next decade, the more severe would be the impacts. The reason for this is that, with time, the U.S. economy would have exploited all the least-cost energy-conservation measures; room for short-term negotiation in case of severe shortages would become progressively smaller with time. In addition, major government supply initiatives are not likely to have impacts before the end of the decade. Thus it is in the best interests of domestic stability that the United States follow a foreign policy that minimizes the probability of future embargo.

**Table 11-4
Relative Impacts of Embargoes in Different Years**

Year	Reduction in Nonenergy Output		Percentage Increase in Domestic Oil Price	Percentage Decrease in Domestic Oil Use	Percentage Increase in Unemployment Rate
	10^9 (1975 $)	Percentage Drop			
Embargo in:					
1981	260	9.1	62	20.0	15
1983	280	13.5	137	18.5	20
1985	320	15.2	187	17.0	22
1990	400	15.6	188	12.8	23

Note: All percentage changes are relative to the base case.

Note

1. For a detailed description of ECONOMY1, see Narasimhan P. Kannan, *Energy, Economic Growth and Equity in the U.S.* (New York: Praeger, 1979); and FOSSIL79, *FOSSIL79 Documentation,* Dartmouth System Dynamics Group (1979).

12 The Role of Electricity in Solving Energy Problems in the Near Term

Thomas H. Lee

This chapter presents some ideas about how to minimize U.S. dependence on foreign oil in the crucial next ten years.

The energy problem in the United States is principally a liquid-fuel problem. The United States used 78 quadrillion BTUs, or 78 quads, of energy in 1978. Of this amount, almost half was oil. Another quarter of the energy supply was natural gas. The balance was provided by coal, nuclear, and hydro, almost entirely in the form of electricity.

How the United States consumes its energy resources tells a more important story (see table 12-1):

> Transportation and feedstocks account for only 32 percent of overall energy consumption; yet they consume over half of U.S. oil—an amount equal to total domestic production.

> Industrial heat and steam consume 22 percent of U.S. energy stock, with oil and gas used about 80 percent of the time in essentially equal amounts.

> Space and water heating, for both commercial and residential purposes, use another 22 percent, more than one-quarter of that amount in the form of oil and slightly less than one-half in the form of natural gas. Residential usage is twice that of the commercial sector.

> The balance of the U.S. energy supply, some 24 percent, is devoted to motors, lighting, refrigeration, and other electrically driven devices.

Transportation and feedstocks and industrial heat and steam represent 55 percent of total U.S. energy consumption and over 75 percent of oil usage.

Looking at the supply side of the equation (table 12-2), U.S. imports are principally 16.5 quads of oil, compared with about 1 quad of gas. Thus, to solve its energy problem, the United States must find substitute sources of energy for oil.

113

Table 12-1
Estimated 1979 Energy Use
(Quadrillion BTU)

	Coal	Petroleum	Gas	Electricity	Total	Percentage
Transportation	—	20	—	—	20	26
Materials	1	3	1	—	5	6
Industrial heat and steam	3	6	7	1	17	22
Residential and commercial space and water heat	—	5	8	4	17	22
Motors, lighting, refrigeration, etc.[a]	—	—	—	19	19	24
	4	34	16	24	78	100

	\multicolumn{4}{c}{Other}		
Electricity	11	3	4	6	24

Source: Based on *DOE Monthly Energy Review* and various end-use energy tables.
[a]Includes motor to drive pumps and fans in heating applications.

The concept to be examined here is a substitutional process that can result in the saving of 50 percent of imported oil within the next ten years. First, consider some of the hard facts. The transportation and petrochemical-feedstock uses are, and will continue for at least the next decade to be, almost exclusively dependent on oil. Electrification of U.S. railroads in a program similar to the interstate-highway program of the 1950s and 1960s

Table 12-2
Domestic Petroleum Sources, Adjusted for Storage and Exports

Petroleum Sources	Year 1973		Year 1979	
	(MBOE/D)	(Percentage)	(MBOE/D)	(Percentage)
Domestic				
Crude oil	9.21	55	8.51	48
Natural-gas liquids	1.74	10	1.66	9
Imported[a]				
Crude oil	3.22	19	5.88	34
Refined products	2.62	16	1.63	9
Total supply	16.8	100	17.7	100

1979 Import-Oil Summary	Year 1979	
	(MBOE/D)	(Percentage)
Crude oil	5.88	78
Gasoline	0.18	2
Distillate fuel oil	0.20	3
Residual fuel oil	1.10	15
Jet fuel	0.07	1
Miscellaneous	0.08	1
Total	7.51	100

Note: No adjustment for heating-value differences of ±10 percent.
[a]87 percent from OPEC sources.

The Role of Electricity

could provide some relief, but not until the late 1980s. The same is true for the electric vehicles that are now under development. Gasohol and hydrocarbon liquid fuels produced from coal, biomass, or natural gas will also be of some help by the end of the decade, and thus deserve accelerated development.

For the crucial ten years ahead, however, even with the beneficial effects of conservation, it may be impossible to reduce significantly the amount of oil required to supply transportation and feedstock requirements without affecting economic growth. These two end uses alone will consume the entire domestic production of oil. Thus, if we are to avoid prolonging heavy U.S. dependence on foreign sources, we must identify ways to free oil from other uses.

A small percentage of electricity, less than 15 percent, is generated by oil. That is not insignificant; it is over 3 quads. With the current drive to convert to coal, some oil will be conserved. But what else can one do to save more oil?

Six quads of oil—roughly 40 percent of U.S. imports—were consumed in the industrial sector for process steam and direct-heat applications. Herein lies a significant opportunity for substitution by either gas, coal, or electricity. Exactly how much can be accomplished in this sector and by what fuel is somewhat uncertain because of our limited knowledge of the relative importance of different end uses. For the moment, assume that for most of these applications, gas can be used to replace oil. This represents 30 percent of the gas produced in the United States. Since the balance between gas production and consumption is very close, this could tax the gas industry or call for more imports of gas. There is increasing evidence that far more nonconventional gas may exist than most forecasts indicate. If this is so, it is simply wonderful. However, it is questionable whether we can increase production sufficiently in the next ten years. Assume that we cannot and, further, that we also do not want to import gas. Where are the opportunities to save some gas so that gas can be used to save oil? There is an opportunity in the commercial and residential sectors, where space and water heating accounted for almost one-quarter of total U.S. energy consumption in 1979. Natural gas accounted for almost half, and over one-quarter, or 5 quads, was in the form of oil. Electricity can be used to save some oil and gas in these applications. This possibility is the subject of a hot debate between the electric industry and the gas industry. Here, however, I will address only the possibilities. For this purpose, I quote my testimony to Congress:

> I should point out that in arguing for substitution of electricity for oil and natural gas, my purpose is not to debate the relative economies of electricity versus oil and gas. I don't believe our liquid fuel crisis is a pure economic problem anyhow. Our national security depends on our ability to reduce the magnitude of this problem. And even more critical is our ability to accomplish that within as short a period of time as possible. The international situation may not allow us the luxury of waiting until the 1990s to solve the

problem with all the long range programs we hear about. Because of this, my purpose is to suggest that in the next ten years there are real opportunities for accomplishing our objectives by shifting energy sources among sectors so that oil will be used only for those applications where we have no choice.

To return to the opportunities in the residential and commercial sectors, in new homes the effort to save oil has already taken place. Half of all homes built over the last five years use electric heat. Over one-quarter of those now under construction are equipped with highly efficient heat pumps. Similar trends are underway in commercial construction.

New construction is not the only way to free oil from space and water heating. Converting older homes to electricity could also help. To predict the savings, we must know how many homes are equipped with warm-air furnaces where the substitution of electricity for oil or gas is the easiest. Census figures indicate that approximately 50 percent of the over 80 million homes in the United States are equipped with warm-air furnaces, with approximately half of these centrally air conditioned. What would be the result if we replaced these warm-air furnaces with heat pumps?

If we assume that in addition to the oil directly saved, the freed natural gas will be used to replace oil either in hot-water or steam heating systems or in industrial applications, then oil usage would be reduced about 3 quads. If we similarly assume that heat pumps could replace 50 percent of the commercial space-heating uses of oil, then oil usage would be reduced an additional 2 quads. If we eliminate the oil used to generate electricity, the saving is 3 quads. (Of course, there is the question of whether all the 3 quads can be saved by coal alone.) The net result is that oil usage can be reduced by 8 quads, or 50 percent of U.S. imports. This is before we even assume significant substitution in the industrial sector. It also does not include potential water-heating or other residential or commercial substitution.

If the situation looks so promising, why are we not doing these things? The first question is: Can the electric industry meet the challenge? In July 1979, the National Electric Reliability Council (NERC) forecast an installed capacity of over 700 GW of generating capacity by 1985. If the load growth between now and 1985 is limited to 4 percent, the need is only 600 GW. An extra 100 GW would then be available. Each GW is equivalent to approximately 24,000 barrels per day. 100 GW is equivalent to 2.5 million barrels per day, or 30 percent of our oil imports. If we look at the 1976 plan (figure 12-1), the difference is another 200 GW. Thus we have two choices: (1) to slow down the construction plants in the pipelines (that is what the industry is doing now); or (2) to accelerate the construction so that we can save 50 percent of our oil imports.

The choice seems clear when explained in economic terms. An additional 100 GW in ten years would cost $100 billion, or $10 billion a year,

The Role of Electricity

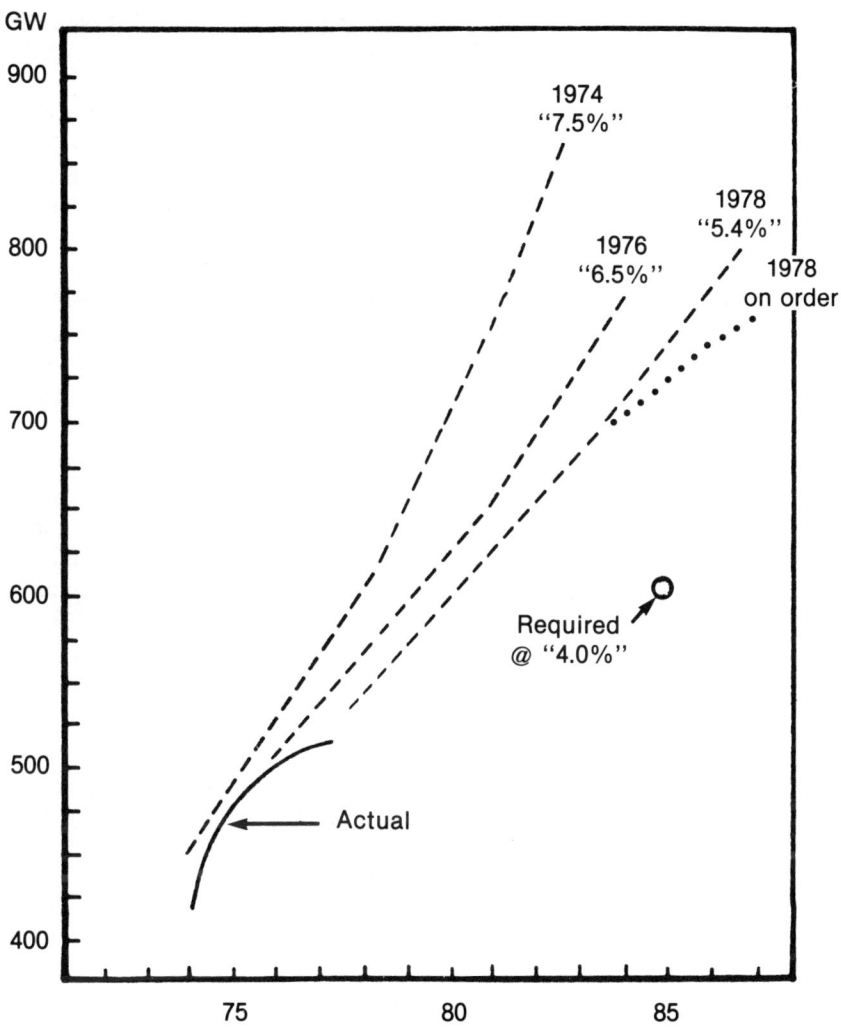

Source: Based on *DOE Monthly Energy Review*, January 1980.

Figure 12-1. Total Installed Capacity, United States

which is equivalent to saving the cost of 1 million barrels of oil per day in one year.

What about the cost for this two-step conversion? To substitute gas for end uses that depend on oil (other than transportation) is a relatively easy task. The cost lies in the residential and commercial sectors and could be in the range of $100-$200 billion. In ten years, this is still a manageable level.

However, we must find ways to manage it. We must ask: What is the relationship between $1 billion spent in the United States and $1 billion spent in OPEC countries?

The second question is: Can the consumer afford to pay for the electricity? This is at the heart of the debate between the gas and the electric industries. I will not exercise judgment on the question of marginal pricing versus average pricing. The historical trend shows that electricity prices have not gone up as fast as oil and gas prices, because fuel prices are only one term in the equation for electricity prices. Capital cost is another term. In an inflationary world, and without inflation-adjusted accounting, the term due to capital cost in electricity prices decreases with time. Of course, one can argue that this is not right. But to return to the original point, I am not arguing about which energy source is cheaper. I am searching for a way to save oil.

The third question is the total cost. If we decide to install heat pumps in the residential and commercial sectors, who will bear the burden of the initial capital cost? The total bill can be a large one, but I believe it is a manageable one. The real question is who will pay. Consumers have a tendency to base buying decisions on first cost. Unless something is done to encourage them to buy, this will be a real obstacle.

For the industrial sectors the problem is somewhat different. If a reasonable return on investment can be demonstrated, and if law permits, industries would probably make these changes. But there are problems in both areas. There has apparently been little work done in looking at different ways to supply energy to different industries. We need a methodology to approach this general subject. We also have problems with our laws. Laws such as the Fuel Use Act do not have saving oil as their underlying objective.

There is another important question: Why should the gas industry be interested in shifting from one sector to another if, in the end, it loses by such a shift. Rate structures and incentives for investment are only two issues that must be addressed. However, despite all the obstacles, the concept proposed deserves careful examination because the next ten years may be the most crucial period. We must ask ourselves how many ways we can think of to make a difference in the 1980s.

13 Relationship between Energy Growth and Economic Growth: The Point of View of the Developing Countries

Marcelo Alonso

Since economic growth is closely related to business growth in a market economy, and one of the essential inputs for economic growth is energy, it is very appropriate that this book, which explores the connection between energy and business, has taken as the first topic to be discussed the relationship between energy growth and economic growth. There is no doubt that economic output demands energy, but these are two complex concepts that cannot be reduced to simple terms. Economic growth implies increased production in a variety of industrial sectors, in a multitude of agricultural activities, and in a series of supporting areas such as transportation and household use—each one having a different energy intensity. Also, not all forms of energy are interchangeable; therefore, the availability of adequate forms of energy to match the demand of the economic sectors is crucial for sustained economic growth. In turn, the availability of energy is determined by the existence of local resources or the capacity to acquire abroad the energy resources required to supplement national production. This complex picture is aggravated by the existence of an extremely versatile primary energy source, oil, which through the marvels of petrochemical technology can be adapted to perform a great variety of tasks and therefore has become the favorite energy source as well as the raw material for many products. However, oil, which is still plentiful, eventually will be exhausted; and long before that it will become scarce and very expensive. On the other hand, the price of oil has gone up by a factor of 15 in the last seven years, and it is impossible to predict with certitude the future trend of oil price, except that it will tend to rise. The most we can hope for at present is that OPEC will reach some kind of formula for indexing oil prices to some combination of industrialized countries' export prices and the value of the U.S. dollar, whatever that might be, thus providing a basis for estimating trends in oil

---

This paper is an expanded and revised version of comments made at the International Chamber of Commerce Conference on Energy held in Lisbon, 3-5 November 1980, to comment on another paper, prepared by Professor H. Houthaker for the same conference and titled, "What Are the Relationships between Energy Growth and Economic Growth?"

prices and minimizing uncertainties. Thus the energy problem in the short term is really an oil problem, which boils down to two considerations:

1. How is it possible to sustain economic growth in a situation of unpredictable and sudden increases in oil price?
2. How is it possible to move away from an oil-based economy as fast as possible, by stringent conservation methods and by appropriate energy substitutions, as well as to increase domestic oil production?

It has often been said that the problem is global and has a profound geopolitical content that at present is almost intractable. However, the problem is so transcendent that it deserves careful analysis. Most of the anlaysis has been macroeconomic in nature and carried out on a country or regional basis, or on countries grouped by certain common features (industrialized or developing, oil exporters or oil importers, and so on). All these analyses provide interesting insights but only give different perspectives of the problem. The addition of all the facets does not give the whole picture, and at the most only certain trends can be identified. To gain a deeper insight into the problem, more detailed analysis is necessary, involving a careful examination of the different economic sectors and the energy inputs required by them, taking into account the circumstances of each country.

Houthaker, in "What Are the Relationships between Energy Growth and Economic Growth?" has given us an overview of the relation between energy growth and economic growth for a group of industrialized countries (ICs), all members of the OECD. I will compare some of his interesting results with those for certain developing countries. First we must agree that there is no such thing as a prototypical developing country (LDC) and that the diversity of situations is tremendous, making generalizations meaningless. It is difficult to offer the same prescriptions for Brazil and India as for Haiti and Burundi.

Nevertheless, from the energy point of view, there are certain important differences between the LDCs and the ICs. One is the large importance of traditional (or noncommercial) fuels in the LDCs (table 13-1), which in some instances may be close to 90 percent (as in Haiti and some African countries), and even in a relatively industrialized developing country such as India may be as high as 48 percent, whereas these fuels are negligible in ICs. Noncommercial fuels are basically immune to oil-price changes and to the worldwide energy situation. However, traditional fuels are used mostly for household purposes and have very little effect in the economic sectors, although their use has several undesirable effects, with clear economic implications such as deforestation and soil erosion. It is important to note that on the average the role of the traditional fuels is declining rapidly because of

Table 13-1
Estimates of Commercial and Noncommercial Fuel Use as a Percentage of Total Energy Requirements

	1967		1972		1977	
	Commercial	Noncommercial	Commercial	Noncommercial	Commercial	Noncommercial
India	72	28	74	26	77	23
Kenya	28	72	31	69	36	64
Egypt	98	2	98	2	98	2
Brazil	57	43	70	30	75	25
Jamaica	91	9	94	6	96	4
Republic of Korea	84	16	91	9	95	5

Source: International Energy Agency, *Workshop on Energy Data of Developing Countries* (1979).
Note: Noncommercial = fuelwood and bagasse.

a multitude of well-known circumstances, particularly industrialization and urbanization. These fuels are rapidly being replaced by oil-based fuels. Industrialization has given rise to a shift from low-energy-intensity agriculture to high-energy-intensity industrial production; and urbanization results in increasing demand for centrally produced services and transportation, for which noncommercial energy is not suitable. For oil-importing LDCs, this places a severe strain on their financial situation, resulting in a steady increase in their external debt (table 13-2). A second difference is in the structure of energy use: in LDCs the household sector uses proportionately more energy and the transport sector less than in ICs, as exemplified in table 13-3 for the United States and India.

Another difference is the higher dependence of the LDCs on oil as a primary energy source, which in some cases is more than 90 percent and on average is around 70 percent, whereas in the ICs it is on average around 48 percent, although the ICs consume about 70 percent of the world's oil production. This is basically the result of the late industrialization of the LDCs, which tended to incorporate a higher proportion of oil-intensive methods of production developed in the ICs; however, most ICs continued using coal to a great extent and can revert to intensive coal use with relative ease. In fact, OECD countries plan to increase the use of coal by three times and nuclear energy by fives times by the year 2020. This circumstance, combined with the much more reduced flexibility of LDCs for changing patterns of energy use and a production structure highly dependent on external parameters, in-

Table 13-2
External Debt of Selected Developing Countries, 1970-1977
(millions of U.S. dollars)

	1970	1977	Percentage Increase, 1970-1977	As a Percentage of GNP	
				1970	1977
Bangladesh	—	2,291	—	—	41.8
India	7,935	14,531	83	14.8	14.7
Sri Lanka	317	787	148	17.1	27.8
Haiti	40	126	215	10.3	10.7
Kenya	313	821	162	20.3	19.7
Egypt	1,639	8,099	394	23.7	69.2
Philippines	630	2,985	374	9.2	14.4
Republic of Korea	1,797	8,472	371	21.5	26.9
Jamaica	154	896	482	11.5	28.7
Brazil	3,405	19,221	464	7.6	11.8

Source: World Bank, "World Development Indicators," *World Development Report 1979* (New York: Oxford University Press, 1979). Reprinted with permission.

Note: Public debt outstanding and disbursed.

Table 13-3
Structure of Energy Use in the United States and India
(percent)

	Household	Industry	Transport	Agriculture	Other
United States	14	30	25	4	27
India	44	30	17	6	3

Source: Professor A.K.N. Reddy, Indian Institute of Science, Bangalore.

troduces an undesirable rigidity in the energy situation of LDCs that is much less pronounced in the ICs.

A fourth and important difference between the ICs and the LDCs is the energy consumption per capita, which in the former is on the order of 4-10 kW per capita, whereas in the latter it is much less than 1 kW per capita, with great variations from country to country and among population sectors within the same country. This, among other things, leaves very little room for conservation and puts great pressure for more energy use as the standard of living improves, particularly that of the poorer people. In fact, commercial energy use has increased lately at a faster rate in LDCs than in ICs (table 13-4). Besides, even if oil prices were kept fixed, the financial requirements to increase the energy per capita in LDCs to the world average of 2 kW/cap, assuming at the same time a more equitable distribution of income, would require an enormous investment. Just for Latin America I have estimated it to be on the order of 3×10^{12}, or about 3×10^4/cap, for which the resources are simply not available, and the economic capability is insufficient.

Last, but not least, the LDCs carry, relative to the ICs and the petroleum-exporting countries (PECs), very little weight in international decisions with respect to energy price and availability. They are simply spectators dragged into the arena where energy battles are fought. They have very little room to maneuver and are highly dependent on policy decisions adopted by the ICs and on financial mechanisms controlled by ICs and, more recently, by PECs.

Table 13-4
Growth in Commercial Energy
(percent)

	1960-1965	1965-1973	1973-1978
ICs	5.0	4.7	1.5
LDCs	−0.1	6.9	7.3

All the foregoing considerations make the analysis of the economic impact of the energy crisis quite different for the LDCs than for the ICs or PECs. Thus whereas in 1980 the PECs had a surplus of 115×10^9, the ICs will experience a deficit of about 50×10^9 and the LDCs, which import oil at a rate of about 5×10^6 Bbl/day, will have a deficit estimated to reach 70×10^9. This deficit exceeds by more than 30 percent the total revenues from their export accounts and is about three × the total external-assistance aid received by LDCs.

Going back to the first point raised—the possibility of sustaining economic growth in a situation of continuous and unpredictable increases in the price of oil—the curious situation is that the LDCs as a group have continued to grow at a faster rate than the ICs which suffered a serious recession during 1974-1975, as illustrated in table 13-5. This, of course, has been accompanied by a much larger inflation rate, as shown in table 13-6. For LDCs average inflation rates rose from about 10 percent to about 30 percent, although with great variations among the LDCs, as seen in table 13-7. Of course, most LDCs have suffered from chronic high rates of inflation; increasing oil prices are just another factor that contribute to the situation—a substantial factor, but not the only one.

One other factor that must be taken into account is the relatively large rate of population growth in LDCs. In fact, the average rate of growth of the GDP per capita in LDCs actually dropped after 1973 from about 3 percent to about 1.5 percent because of a large rate of population growth compared with the rate of economic growth.

How has it been possible to sustain economic growth under such conditions? On the one hand, some countries, such as Korea, were able to increase their production, in some instances doubling the value of their exports. In most cases, however, economic growth has been achieved by borrowing money for financing the energy these countries had to import. This, of course, has deteriorated the current accounts of those countries, particularly because many of the loans are short term; but it has allowed them to finance the energy required by their development plans (table 13-8). Of course, those countries that could do neither one thing nor the other suffered the most. Recently the World Bank asked for 9×10^9 to help finance the external debt of LDCs resulting from higher oil prices.

Table 13-5
Growth in GDP
(percentage annual rate)

	1967-1972	1973	1974	1975	1976	1977
ICs	4.6	6.1	0.1	−0.9	5.4	3.7

Energy Growth and Economic Growth

Table 13-6
Rate of Inflation
(percent)

	1967-1972	1973	1974	1975	1976	1977
ICs	4.1	7.3	11.9	10.9	7.2	6.9
LDCs	10.1	22.1	33.0	32.9	32.3	31.5

The relationship between energy use and economic growth differs greatly from one country to another, as illustrated in table 13-9. It is very difficult to reach any conclusion from that data unless one examines in more detail the structure of energy demand and of the economic sectors. One can say, however, that as long as energy has been available, the countries have been able to continue to grow, although at the price of incurring ever-increasing debts. Table 13-2 gave the rate of growth of the external debt of several LDCs. Again, although there are great variations from one country to another, the general trend is clearly upward. The balance-of-payments deficit has nearly tripled since 1973.

Has there been a real effort to improve the energy intensities in LDCs? Table 13-10 (which is similar to Houthaker's table II, reproduced here as table 13-11 for convenience) presents the situation for a series of LDCs. Again, it is very difficult to draw meaningful conclusions, except that ap-

Table 13-7
Inflation Rates in Developing Countries

	1979	1980	1981 (Expected)
Mexico[a]	20	28	24
Brazil	52	97	74
Venezuela[a]	13	24	18
Argentina	160	100	85
Colombia[a]	24	26	23
Peru	67	70	70
Tanzania	9	7	6
Ghana	20	14	13
Kenya	7	6	6
Nigeria[a]	15	13	14
Zaire	22	18	19
Korea	18	20	12
Taiwan	6	11	8
Singapore	4	10	12
Philippines	18	20	16
Thailand	15	14	12
Indonesia[a]	25	18	14

[a]Oil exporters.

Table 13-8
Current Account Balances for Selected Developing Countries
(millions of U.S. dollars)

	1970	1973	1974	1977	1978
Bangladesh	—	−235.5	−475.1	−287.9	−386.9
India	−392	−529	+1,207	NA	NA
Sri Lanka	−58.5	−25.2	−135.9	+136.2	−63.8
Haiti	+1.7	−1.4	−22.9	−52.4	NA
Kenya	−490	−126	−307.9	+56.8	−538.8
Egypt	−154	+77	−327	−1,201	−1,222
Philippines	−48	+474	−207	−830	−1,157
Republic of Korea	−623	−307	−2,027	+7	−1,141
Jamaica	−152.9	−247.6	−90.9	−68	−137.9
Brazil	−561	−1,759	−7,180	−5,117	−7,034

Sources: (1970-1974): International Monetary Fund, *International Financial Statistics* (December 1977); (1977-1978): International Monetary Fund, *International Financial Statistics* (December 1979).

parently most oil-producing LDCs have not made special efforts or have not been able to improve their energy intensities, whereas the oil-importing LDCs show many fluctuations, suggesting either lack of national policies or an intrinsic impossibility of making improvements. In the first place, in absolute values the energy intensities of LDCs vary tremendously because of the diversity of their productive structures, as exemplified by the numbers in parentheses for 1973. In the second place, the possibilities of the countries improving their energy intensities in the short term are also different for each country, depending on their size, the nature of the demand, the technologies to which they have access, and government policies.

This brings us to our last question: To what extent has economic growth been more influenced by domestic policies than by the energy situation? Obviously, as was said before, to the extent that domestic policies have been allowed to trade energy for external debt, or that it has been possible to increase exports, economic growth has been sustained; but anti-inflationary policies, sometimes required by external agencies such as the International Monetary Fund (IMF), and other times dictated by internal reasons, may seriously affect economic growth. Since inflation is a complex phenomenon that couples both ICs and LDCs, through a push-pull effect, but affects the LDCs more seriously because the prices of many of their products cannot be adjusted to reflect inflation (differentiating manufactured products with great energy value added from agricultural and mineral products with relatively small energy valued added), the anti-inflationary policies are by necessity of limited effect in LDCs. In this respect Houthaker claims that energy pricing by governments can be an effective instrument in reducing energy consumption. Although this is important in ICs where energy per

Table 13-9
Increase in Energy Consumption and GDP

	Commercial Energy Consumption (Percentage Increase per Year)		GDP (Percentage Increase per Year)		Energy Intensity (Percentage Increase)	
	1960-1973	1973-1976	1960-1973	1973-1976	1960-1973	1973-1976
Algeria[a]	8.0	20.0	3.5	3.8	4.5	16.2
Egypt[a]	2.9	17.1	4.7	6.8	−1.8	10.3
Kenya	4.0	1.6	5.5	3.2	−1.5	−1.6
Nigeria[a]	9.5	7.6	5.5	6.4	4.0	1.2
Brazil	6.3	8.3	7.6	8.1	−1.3	0.2
Colombia[a]	5.4	3.6	5.6	4.8	−0.2	−0.8
Jamaica	14.3	0	4.7	−2.4	9.6	2.4
Mexico[a]	6.8	5.3	6.8	3.9	0	1.4
Venezuela[a]	6.4	6.2	5.7	6.2	0.7	0
India	4.6	6.3	3.7	3.2	0.6	3.1
Indonesia[a]	4.2	13.9	5.0	6.5	−0.8	7.4
Korea	12.7	6.6	9.2	10.8	3.5	−4.8
Philippines	9.1	5.1	5.4	6.5	3.7	−1.4
Thailand	16.6	1.8	7.7	6.8	8.9	−5.0
Portugal	8.7	0.3	6.9	1.9	1.8	−1.5
Turkey	10.0	8.9	5.9	8.7	4.1	0.2

Source: Based on data from L. Gordon, *Energy Planning and Economic Development*, Resources for the Future (February 1979).
[a]Oil exporters.

capita is high, and although it might work in LDCs, it cannot be a dominant factor because of the need of LDCs to increase their use of energy per capita, as mentioned earlier.

To conclude this brief and incomplete analysis, I will refer to the second point raised at the beginning: How can the LDCs move effectively away from an oil-based economy? We have already indicated that conservation can play only a limited role, whereas the energy alternatives presently available cannot have a very important effect in the short and medium terms (mostly for financial and technical reasons). Increased use of solar energy, biomass and biogas, and other nonconventional energy sources will essentially replace traditional fuels, whereas coal and nuclear energy are options that at present can be exercised by only a few LDCs, although in the long term they might become more important for most LDCs. All other energy alternatives are of much longer-term effect. Thus, although I am confident that the energy problem will eventually be resolved on a global scale, the problem will remain with us for the short term, or at least until the year 2000; and in this transition period the problem (particularly for the LDCs), rather than technology and resource availability, is basically a geopolitical and financial one. No simple solutions exist now, but both PECs and ICs bear the primary responsibility of achieving them.

Table 13-10
Energy/Output Ratios
(1973 = 100 percent)

	1960	1961	1970	1973	1976
Algeria[a]	58	63	77	100 (549)	138
Egypt[a]	124	112	76	100 (1,370)	131
Kenya	121	110	90	100 (666)	95
Nigeria[a]	62	84	72	100 (198)	103
Brazil	96	98	100	100 (588)	100
Colombia[a]	103	103	104	100 (1,248)	97
Jamaica	31	55	70	100 (1,372)	108
Mexico[a]	100	83	97	100 (914)	104
Venezuela	91	90	92	100 (1,201)	100
India	89	101	93	100 (1,420)	110
Indonesia[a]	111	99	89	100 (764)	122
Korea	67	95	115	100 (1,873)	77
Philippines	64	85	107	100 (883)	96
Thailand	36	60	89	100 (972)	86
Portugal	80	81	83	100 (659)	95
Turkey	61	77	88	100 (759)	101

Source: Based on data from J. Dunkerley et al., *Assessment of Energy Demand for Selected Developing Counties*, Resources for the Future (August 1980).
Note: Value in parenthesis for 1973 in TOE/M$.
[a]Oil exporter.

Table 13-11
Ratio of Total Energy Requirements to Real GDP by Country (OECD)
(Index numbers, 1975 = 100)

	Canada	United States	Japan	France	Germany	Italy	United Kingdom
1960	101.8	99.6	104.1	113.3	105.2	76.8	122.0
1970	98.5	105.6	108.9	109.5	107.6	103.6	116.0
1972	101.5	104.6	104.0	107.0	106.5	105.8	111.5
1973	97.1	102.3	102.4	113.8	107.9	104.5	107.6
1974	99.2	101.9	103.7	106.6	113.1	101.2	104.6
1975	100.0	100.0	100.0	100.0	100.0	100.0	100.0
1976	96.4	100.4	97.1	101.0	103.1	101.0	98.0
1977	96.7	98.1	93.3	99.1	100.0	100.8	99.2
1978	97.4	96.1	90.3	101.8	100.6	97.3	96.2

Source: H. Houthaker, "What Are the Relationships between Energy Growth and Economic Growth?" Reproduced with permission from H. Houthaker, "Relationships between Energy Growth and Economic Growth," *World Economic Yearbook 1981* (Paris: ICC Services, 1981).

Appendix 13A

In describing the relation between energy use and economic output, two different kinds of coefficients are normally employed. One is the *energy intensity*, *I*, defined by

$$I = \frac{E}{G}, \qquad (13A\text{-}1)$$

where E is the total energy used and G is the gross domestic product (GDP). The energy intensity gives the energy used per unit of GDP. Its numerical value depends on the units chosen for E and G. It is of the order of unity when E is expressed in W and G in \$/year. If the relation between E and G were truly linear, I would be a constant, both in time and for all economic situations. However, that is not the case, since for a given GDP the energy used depends on the structure of the productive sector, including the type of economic activities and the efficiency of the processes employed, as well as the social demand for energy. In any case, conventional wisdom suggests that the energy intensity varies with GDP, as illustrated in figure 13A-1. In the period of emerging industrialization, I increases because of the trend toward more energy-intensive activities; but after a certain level of development, I begins to decrease because the use of more efficient processes allows the use of less energy per unit of output. The ICs are supposedly in the downward part of the curve, whereas the LDCs are still in the upward section. It can be presumed, however, that by using development shortcuts such as early introduction of energy-efficient processes, the LDCs can evolve, as indicated by the dotted line, and rapidly join the IC curve.

By differentiating equation 13A.1 logarithmically, one obtains

$$\frac{\Delta I}{I} = \frac{\Delta E}{E} - \frac{\Delta G}{G}. \qquad (13A\text{-}2)$$

This equation, which has the advantage of being independent of the units chosen for E and G, relates the fractional change in energy intensity to the fractional changes in energy used and in GDP. It is more interesting because it reflects the dynamics of the situation. Obviously one objective of any energy policy is to make $\Delta I/I$ a negative quantity. For ICs the ratio $\Delta I/I$ is negative, whereas for most LDCs it is positive if we analyze the energy trends over the last seven or eight years.

Another coefficient is *energy elasticity*. It can be assumed that the fractional change in energy use is related to the fractional changes in economic output and in energy price by the relation

$$\frac{\Delta E}{E} = \alpha \frac{\Delta G}{G} + \beta \frac{\Delta P}{P}, \qquad (13A\text{-}3)$$

where α is the *income energy elasticity*, and β is the *price energy elasticity*. Both coefficients are independent of the units used for the quantities in equation 13A-3. The equation is an oversimplification because there may be other factors affecting energy use. Also, E is an aggregate of several energies (oil, coal, gas, nuclear, hydro, solar, and so on), each having its own price structure; and G depends on the correlation between industrial structure and energy mix. Thus, rather than one equation we should have a set of equations, one for each type of energy, with different sets of elasticities:

$$\frac{\Delta E_i}{E_i} = \sum_k \alpha_{ik} \frac{\Delta G_k}{G_k} + \sum_j \beta_{ij} \frac{\Delta P_j}{P_j}, \qquad (13A\text{-}4)$$

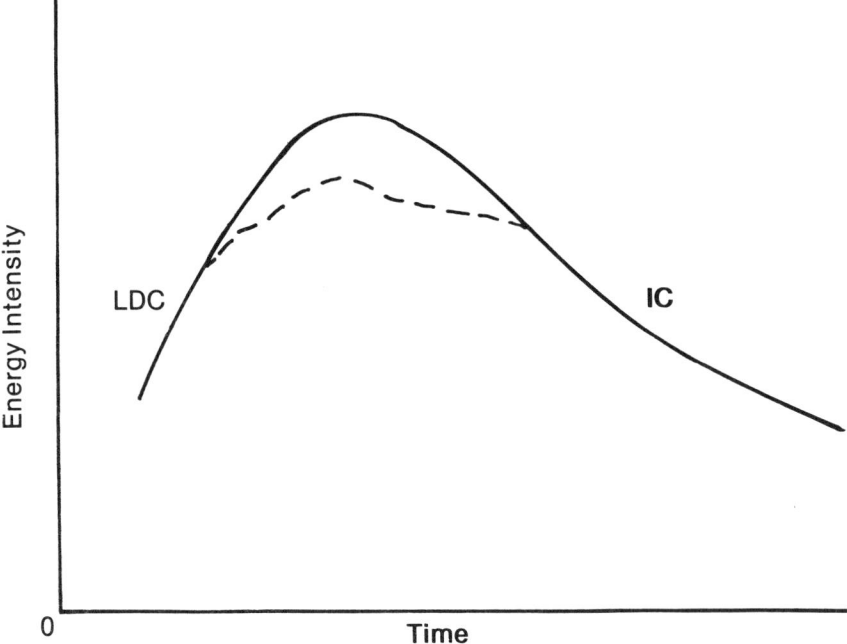

Figure 13A-1. Energy Intensity as a Function of Time for LDCs and ICs, Indicating Possible Development Shortcuts during Transition (Schematic)

where the subindexes i, j refer to the different kinds of energy, and k refers to the different economic sectors. Even so, these equations are imperfect to the extent that they do not incorporate other factors that are not of direct economic nature. For these reasons α and β cannot be considered constant coefficients and depend on several factors. In any case, β should be negative, since an increase in the price of energy should result in a decrease in the use of energy. There may, however, be exceptions to this rule; Korea is an example in recent years.

For the ICs the value of α is less than one, whereas in general α is larger than one for LDCs. Similarly, the absolute value of β is larger for ICs than for LDCs.

References

Cooper, C.L., and Koshel, P.P. 1979. *Energy and Development.* Institute for Energy Analysis.

Dunkerley, Joy. 1979a. *Adjustment to Higher Oil Prices.* Resources for the Future (June).

———. 1979b. *Trends in Energy Consumption and Economic Growth.* Resources for the Future (June).

———. 1980. *Estimation of Demand and Conservation—The Developing Countries.* Resources for the Future (June).

Dunkerley, Joy, et al. 1980. *Assessment of Energy Demand for Selected Developing Countries.* Resources for the Future (August).

Fallen-Bailey, D.G., and Byer, T.A. 1979. *Energy Options and Policy Issues in Developing Countries.* World Bank (August).

The Geopolitics of Oil. 1980. Washington, D.C.: U.S. Government Printing Office.

Gordon, L. 1979a. *Energy Planning and Economic Development.* Resources for the Future (February).

———. 1979b. *Energy Implications of Alternative Development Patterns.* Resources for the Future (February).

———. 1979c. *Energy Issues in Developing Countries.* Resources for the Future (June).

———. 1979d. *International Energy Arrangements and the Developing Countries.* Resources for the Future (July).

Hughart, D. 1979. *Prospects for Traditional and Nonconventional Energy Sources in Developing Countries.* World Bank (July).

Parikh, J.K. 1978. *Energy and Development.* World Bank (August).

Patel, S.J. 1978. *Energy Policies and Collective Self-Reliance.* Vienna Institute for Development.

Resources for the Future. Forthcoming. *Energy and Development: Crisis and Transition.*

Trehan, R.K.; Leigh, J.C.; and Park, W.R. 1979. *Energy and Development.* Mitre Corporation (July).

World Bank. 1980. *Energy in the Developing Countries* (August).

14 Alternative Energy Options for Puerto Rico

Eduardo López-Ballori

Introduction

Puerto Rico is a complex society that in many ways is very difficult to classify as either a developing or a developed country. It could even be said to be an *overdeveloped* country, in the sense used by many environmentalists. Furthermore, its socioeconomic and political structure is in a rapid state of flux, which creates an additional uncertainty in attempting any future planning for the island. In comparison with many of the developing countries that, like Puerto Rico, are almost totally dependent on imported fossil fuels to satisfy their energy needs, Puerto Rico is considerably more industrialized and enjoys a substantially higher standard of living than most of these countries. Another interesting fact is that Puerto Rico's per capita energy consumption (equivalent to 18 barrels of oil per year) is greater than that of some of the developed countries—for example, Italy.

It is well known that the high level of economic and social growth attained by Puerto Rico during the past three decades was made possible by the availability of low-cost energy in the form of imported oil from countries that now belong to the OPEC cartel. In 1973 Puerto Rico, based on its absolute consumption of oil, was number 23 in the list of the 126 petroleum-consuming countries of the world represented at the United Nations. This fact should be looked at in light of a population of 3.1 million and a territorial extension of only 3,400 square miles. In 1975 the island was among the first ten countries in the world in terms of the density of primary energy use (energy per square mile of active land). Not surprisingly, the enormous rise in the cost of this energy source during the past decade has produced economic dislocations in Puerto Rico and a serious challenge to its plans for the continuation of its socioeconomic development.

Puerto Rico and Other Developing Countries

Puerto Rico shares with many developing countries some features and structural problems, such as the following:

Features

1. Puerto Rico does not posssess any proved desposits of oil, coal, or gas. The island is currently exploring for potential oil and natural-gas reserves off its northern and southern coasts, but it cannot rely on the possibility of a find as an answer to its energy needs.
2. Puerto Rico depends on petroleum for all but 1 percent of its energy needs, a dependence far greater than that of typical developed countries (the United States depends on oil for 40 percent of its energy needs).
3. In addition to importing almost all the energy it consumes, Puerto Rico imports most of its petroleum from OPEC countries. In the last three years, however, there has been a rising tendency to bring crude oil from non-OPEC countries, particularly from the United States. In 1979 17 percent of the oil imports came from the United States.
4. The island does not have adequate storage facilities for crude oil and refined products, which results in an inability to store oil as a protection against a severe interruption of supplies.

Structural Problems

1. During the last four decades of economic growth, Puerto Rico has suffered a reduction in agricultural activitiy that has led to an almost unbearable situation in which most foodstuffs are imported.
2. Unemployment and underemployment remain high, problems that are further compounded by a relatively high rate of population growth.
3. The healthy survival of many of the island's ecosystems is seriously threatened by the excessive loads imposed by some economic activities and by overpopulation.
4. Uncontrollable rural-urban migration has been detrimental to rural development and has practically converted the island into a *megalopolis*.
5. Patterns of consumption of a developed society have emerged in Puerto Rico with no corresponding increase in the productive capacity necessary to generate the consumption items demanded by the people. Enclosed air-conditioned buildings and residences, gas-guzzling cars, and a multitude of impermanent power-consuming devices aggravates the precarious energy and environmental situations.

Puerto Rico's Energy Structure

To understand the energy structure that has evolved in Puerto Rico, one must keep in mind the status of the island as a developed-developing country. There are several essential facts describing the energy structure.

First, unlike most developing countries, Puerto Rico has a fully integrated electrical network that extends throughout the island. This electric sector, almost fully dependent on fuels derived from petroleum, consumes 40 percent of the island's total oil consumption. The electrical system represents a large capital investment of more than $1 billion by the government of Puerto Rico. This large investment has resulted in a widespread use of electricity by all the economic sectors. Petroleum-price increases during the last decade have led to substantial cost increases for individual, commercial, and industrial consumers who are paying between 30 and 40 percent more for electricity than their counterparts in the Gulf Coast states. The existence of the large petroleum-based electric sector sets considerable limits on the time required for a transition to a postpetroleum era.

Second, the transportation sector, made up mainly of private automobiles, has become a significant oil user (26 percent of all oil resources used in the island in 1979). The significant use of private vehicles (Puerto Rico ranks sixth in per capita car ownership in the world) is not related to affluence, as in developed countries, but is mostly related to the lack of a public-transportation system responsive to the needs of a rapidly expanding suburban population.

Finally, the predominance of energy-intensive industries is a salient feature of the island industrial sector, which is the principal direct consumer of oil (38 percent of total oil consumption). The refinery and petrochemical complexes constitute the most important items in the industrial sector of the island, and they depend exclusively on oil and its by-products. Understandably, the dramatic increase in price of OPEC oil has severely affected the financial health of these industries, whose initial establishment and prosperity depended crucially on the price differential between OPEC oil and mainland oil. At present, Puerto Rico is trying to keep this industry afloat and to restore its capability for the establishment of downstream industries that hold the highest potential in terms of jobs and income.

Two Basic Questions

The energy crisis in Puerto Rico has so far been one of cost and not of availability of oil. This, of course, does not preclude the occurrence of an oil shortage in the future. Therefore, a strategy must be developed for transition to an energy system that does not depend unduly on imported oil. The goals to be pursued in developing such a strategy are:

1. a limited growth in energy consumption compatible with the degree of economic growth necessary for the continued welfare of Puerto Rico;
2. a gradual diversification of energy sources to improve the resilience of the energy system to withstand disruptions;

3. an increase of the economic productivity of the energy used in all economic sectors;
4. the attainment of the highest level of energy self-sufficiency through the application of new technologies based on renewable energy resources.

To accomplish these goals, the following two basic questions must be considered:

1. What are the resources and technological options available that are also adequate to achieve the desired goals?
2. What kind of feasible responses (decisions, institutional changes, programs) are needed to go from where we are now to where we want to be at a specific time in the future?

The rest of this chapter is devoted to these two basic questions, keeping in mind the possible relevance or irrelevance of Puerto Rico's efforts in the context of developing countries.

Energy Options

The search for energy options for Puerto Rico has demonstrated, as everywhere else, that there is no *single* option with the capability to replace imported oil in the foreseeable future. For this reason, to achieve self-sufficiency, a combination of options must be considered. These would have to be used in a mix that would minimize their total systemic costs (economical, social, and environmental). More specifically, they should comply with the following criteria:

1. The technologies should be based on renewable or virtually inexhaustible sources. This is a sine qua non for survival in the twenty-first century.
2. They should have minimum environmental impacts and should in no way overburden the limited assimilative capacity of the island environment.
3. They should be technologies whose scale and relative simplicity permit them to be adaptable to local conditions and to be controlled by the available human and technical resources of Puerto Rico.
4. They should be technologies that can be harmonized with the use of land (the island's most limited resource) for other purposes.
5. They should be options with the highest possible job-creation capability.
6. They should be cost effective and help to improve as much as possible the capital-generation efficiency of the economy. That is, their capital demand should be carefully assessed so as to prevent overdrawing Puerto Rico's investment capacity.
7. Finally, attention must be given to the thermodynamic matching between the energy source and the end uses of the energy so as to mini-

Alternative Energy Options for Puerto Rico

mize waste and negative environmental impacts and to achieve the highest net energy benefits.

As is also the case with many other developing countries with similar geographical and climatological conditions, Puerto Rico's tropical location permits the consideration of a wide range of alternative energy options that meet the previously specified criteria. More specifically:

1. There is a high incidence of solar radiation—5.5 kilowatt-hours per day per square meter, with only slight seasonal and spatial variations, in an ambient temperature between the high seventies and the eighties, which is favorable to the operation of solar devices during the whole year.
2. Ocean-temperature differences of 20°C between warm surface and deep ocean water exist year round within one mile off shore.
3. The prevailing trade winds, enhanced by local effects on the north and northeast sides of the island, provide relatively high and stable velocity winds on the order of 10 miles per hour.
4. Rainfall is relatively abundant and, together with soil conditions, permits dramatic records of biomass production, particularly of sugarcane (30 dry tons per acre per year) and of other tropical grasses such as Napier grass.
5. These resources exist in a relatively benign environment, completely devoid of icing problems and earthquake, and very infrequently affected by hurricanes.
6. These natural energy resources also exist in close physical proximity to one another, facilitating a systems approach for the search and analysis of optimum energy-option mixes.

These conditions make Puerto Rico one of the few areas of the world in which the economic feasibility of various solar-energy options could be proved sooner than in most other places, a factor that in turn will accelerate their commercialization. The options that have the highest priorities are the following:

1. *Direct solar-conversion technologies:* These include solar water heating, solar steam, solar industrial-process heat, and passive solar cooling. Of lower priority in the shorter term, but with long-term importance, are solar direct-cooling technologies and photovoltaic systems. Unexpected advances could permit a faster rate of application than envisioned at present.
2. *Biomass technologies:* The most important biomass sources for the island are sugar cane and other tropical grasses and solid wastes. The following conversion processes have been identified for these biomass sources:

a. direct combustion as an immediate boiler fuel;
b. anaerobic digestion yielding methane and other gases;
c. fermentation yielding ethanol, ethylene, aldehydes, and other chemical feedstocks.

It can be seen that biomass provides fluid fuels that today are derived from petroleum and for which finding substitutions is perhaps the most crucial problem to be solved in the path to energy self-sufficiency.

3. *Indirect solar sources:* These include ocean thermal energy conversion (OTEC), wind energy, and hydroelectricity. At present, with the exception of hydroelectricity, there exist uncertainties and risks associated with both ocean thermal energy and wind energy that suggest a cautious approach to both. At he same time, the favorable conditions offered by the island for these technologies are a stimulus to research and development work on OTEC and studies that would identify favorable wind sites, provide wind-velocity data, and analyze commercialization problems associated with wind energy.

The full energy impact of these technologies, their time framework, and the trade-offs involved are very difficult to assess. This is one of our most serious concerns.

At present, biomass energy (based on the sugar-cane industry and solid wastes) and solar water heating appear to be the most important renewable options for the immediate future. By the end of the decade these technologies could displace perhaps as much as 10 percent of Puerto Rico's current fuel consumption. In hot, humid tropical countries such as Puerto Rico, sugar cane has been shown to be the plant that provides the most efficient conversion of solar energy. Using existing varieties and technologies, grown so as to maximize fiber content instead of sucrose, it is possible to increase sugar-cane yield to around 30-35 dry tons per acre per year. A recent study of the net energy balance shows that energy cane can produce almost ten times as much energy as is consumed in producing it, and that this energy payback can be improved in the future. This will result in a net energy output of about 70 barrels of oil per acre per year. (The energy in a ton of dried biomass is equivalent to that contained in 2.5 barrels of oil). Equally important is the existence of a high level of local expertise to produce the sugar cane as an energy crop efficiently and economically, thus eliminating constraints usually encountered in new technologies.

The biomass option will probably never attain its full potential since it wil demand a substantial portion of the available agricultural land, which otherwise would be used for food production. This trade-off between energy and food is a more acute problem for the island than for developing countries with larger extensions of land, such as Brazil. Nevertheless, if self-sufficiency is to be attained, it will be necessary to find substitutes for oil that can be converted to fluid fuels with a high energy content and that can be produced in large amounts. Biomass-derived fuels are the best choice to meet these criteria.

Solar water heating also offers significant energy potential and is already economically competitive with electric-resistance heating. Collectors for the Puerto Rican climate can be simpler and less expensive than those used in most other areas of the developed countries. The key problem here is how to accelerate their production through use of government incentives and other means.

Puerto Rico's Responses to the Challenge of Transition to the Postpetroleum Era

The government of Puerto Rico has developed a conservation program designed to reduce energy consumption in those sectors with the best potential for reduction. Conservation through technical means such as a new building code and new lighting-efficiency standards in existing buildings are the island's most effective near-term solutions, with significant long-term impacts. These measures are being reinforced by an education campaign through the mass media and other selected means such as community groups and the school system. Such "technical fixes," as well as life-style modifications, are feasible and necessary if Puerto Rico is to achieve its goal of self-sufficiency.

An alternative-energy program has been gradually emerging to stimulate research, development, commercialization, and education in alternative energy options. As a means to provide impetus to this program we have obtained from our legislature a $5 million authorization to be used as matching funds for research and development projects. Initial work has already been made, particularly in the field of biomass. Several bills have been passed aimed at encouraging the use of solar energy for water-heating purposes and the establishment of a solar industry on the island. Active work is being done to develop an adequate mass-transportation system to reduce Puerto Rico's dependence on the private automobile.

Finally, Puerto Rico intends to use federal financial incentives for alternative fuels as much as possible, and the government of Puerto Rico will seek a mutually beneficial partnership with private industry and the federal government to advance alternative energy technologies.

The transition from the present energy system to a more self-sufficient one will take time. How long it will take depends crucially on decisions currently being made. These decisions depend on the existence of a clear awareness of the energy future we want to create. Given the time to develop and commercialize the solar technologies, it is necessary that during the transition period transitional fuels, mostly coal, should be used. A decision has already been made to install a coal plant of 450 MW in Puerto Rico, which is expected to be on line by 1984. Another similar plant is expected to be operative by 1988. In view of the large coal reserves in the United States and in parts of South America, it is expected that coal importation would not be difficult. The coal plant will also promote a process of diversification of energy sources, which is one of the island's long term goals.

Conclusions

Throughout history different energy sources have assumed the primary role. The last fifty years might be called the petroleum era. We are now approaching the end of this era. Transitional periods have always been by nature uncertain and risky. But this does not inhibit human capacity to solve the crisis. Puerto Rico hopes to overcome the current crisis by establishing a self-sufficient energy system through the development of existing natural resources. In spite of its unique circumstances, Puerto Rico's successes and failures can and may be used by other developing and developed countries in pursuit of their own goals of self-sufficiency.

15 Near-Term Feasibility of Candor Fusion Reactors

Bruno Coppi

The term *Candor fusion reactors* refers to those that employ nonradioactive fuels and that involve primarily reactions that do not produce neutrons. On the basis of existing experimental evidence and a recently developed theory on the macroscopic stability of magnetically confined plasmas, it can be shown that experiments to test the ignition conditions of deuterium/helium-3 reactors (which fall into the category of candor reactors) can be feasible within the limits of present-day technologies.

The same type of toroidal experimental devices can also be employed to test the ignition conditions for so-called deuterium partially catalyzed reactors, which can be used to produce the needed helium-3 fuel.

Introduction

A variety of studies on the long-term development of economical energy sources do not predict an effective role for fusion reactors until well into the twenty-first century.[1] Although one may agree with the long-term nature of these prospects, it should be pointed out that this derives only in part from the scientific or technical difficulties that have surfaced during the relevant research efforts carried out so far; rather, it is the product of the benign neglect with which this field of research has been treated, even when substantial financial support has been made available. In fact, with the exception of a few notable cases, the present trend of the fusion program has been to rely on a few large-scale experiments, which will require at least a decade from their inception to their completion, making it difficult to see a rapidly converging time scale leading to a useful energy source.

Given this long-term time frame, fusion research should attempt to deliver a system that will realize to the best extent possible the full potential of a fusion burner. Such a system is represented by a so-called candor reactor, which does not require radioactive fuels and in which the production of high-energy neutrons is minimal. In particular, a deuterium/helium-3 reactor can be considered the prototype of a system of this kind, the primary relevant reaction being:

$$D + He^3 \to He^4 \ (3.67 \text{ MeV}) + p(14.67 \text{ MeV}),$$

whose products are α-particles and protons. These charged particles can be contained by appropriate magnetic-field configurations. Thus their energy is available to heat the plasma and for possible direct conversion to electrical power, whereas the energy with which they impinge on the first wall of the plasma chamber can be controlled in order to minimize their damage.

Conversely, the neutron energy, constituting 80 percent of the total energy released in the deuterium/tritium reaction

$$D + T \to He^4 \ (3.5 \text{ MeV}) + n(14.1 \text{ MeV}),$$

is neither available to heat the plasma nor available for possible direct conversion to electrical energy. In addition, energetic neutrons pose severe materials problems for the first wall of the chamber and the lithium blanket that must be incorporated in a D-T reactor in order both to produce tritium and to shield the magnets. We also note that a D-T reactor lends itself to use as an intense neutron source for the breeding of fissile fuels. On the positive side, this can provide a short-term demonstration of one of the possible practical uses of fusion power and can speed up, for the benefit of fusion research, the acquisition of industrial technologies that have marked the evolution of fission reactors. On the other hand, the development of the fusion breeder may unnecessarily extend in time the use of fission reactors and may increase the dissemination of technologies and materials that can be diverted into undesirable weapons development.

The D-He³ Test Reactor

Until recently the realization of a D-He³ or a D-D burning reactor had been considered a goal to be achieved in the next century, since a near-term experiment to test the possibility of igniting these kinds of plasmas could not be foreseen. However, recent experimental observations and new theoretical developments have led to the identification of a class of experiments that can lead to the realization of ignition conditions for a D-He³ system on the basis of present-day technology. In fact, in order to achieve this goal, it is necessary to have a plasma confinement configuration that can attain, for instance:

1. values of $n_0 \tau_E$ (n_0 being the peak plasma density and τ_E the energy replacement time) higher than 5×10^{14} sec/cm³;
2. values of $<\beta> = 8\pi <p>/<B^2>$ ($<p>$ being the average plasma pressure and $<B^2>/8\pi$ the average magnetic pressure of the confining magnetic field) higher than 10 percent;

3. plasma currrents around 5 MA or higher, in order to generate the magnetic fields needed to confine the 14.7 MeV protons produced;
4. average plasma temperatures around 30 keV or higher.

These objectives can be achieved simultaneously in an axisymmetric toroidal configuration (see, for instance, figure 15-1) in which:

> The first goal is pursued on the basis of currently known scalings for the plasma thermal conductivity that exhibit a favorable dependence of τ_E on n. Thus it is proposed that peak particle densities, exceeding $10^{15} cm^{-3}$, can be obtained in a configuration with sufficiently high magnetic fields (typically 120 kG) having an adequate area of its transverse cross-section in order to meet the desired $n_0 \tau_E$ criterion. Well-confined plasmas with peak density values higher than $10^{15} cm^{-3}$ have, in fact, been produced in the Alcator device at MIT.

> The second goal is pursued by adopting a combination of magnetic and geometric parameters, such as the torus aspect ratio, in such a way that during the heating cycle the plasma is maintained in one of the macroscopically stable regimes that have been identified in recent theoretical developments.[2]

> The adoption of high-magnetic-field technologies will make it possible to induce plasma currents exceeding 5 MA without violating any of the known criteria against macroscopic instabilities.

> The fourth goal could be achieved by adopting an RF heating system to supplement ohmic heating in order to bring a deuterium-tritium plasma

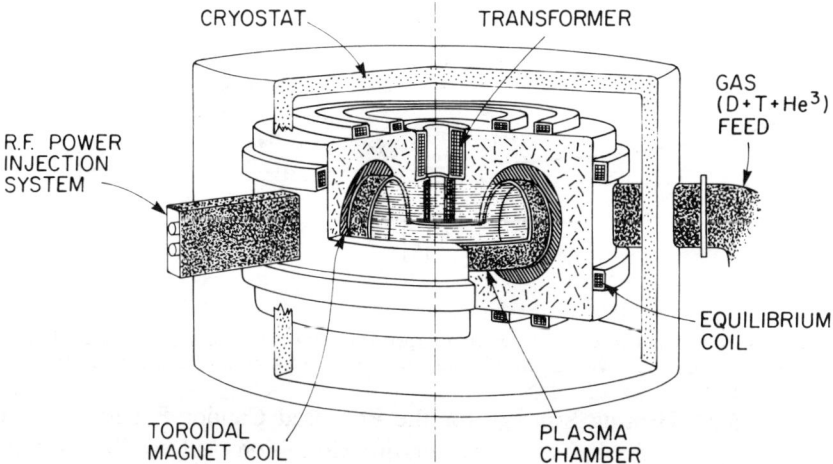

Figure 15-1. Main Components of the Proposed Candor Experiments

to ignition conditions. Thus tritium is used as a "match" to raise the plasma temperature and, as this temperature increases, is gradually replaced by He³ (see figure 15-2). The most convenient frequency for the auxiliary heating system appears to be that corresponding to the first harmonic of the cyclotron frequency of He³ and, at the same time, to the second harmonic of the cyclotron frequency of tritium.

Note that the effectiveness of ion-cyclotron heating in plasmas with two species of ions has been well demonstrated in several experiments carried out on the most advanced existing toroidal devices.

The usually known ignition conditions, based on the assumptions that (1) the distributions of all components of the background plasma are Maxwellian, and (2) the slowing down of the charged fusion-reaction products is

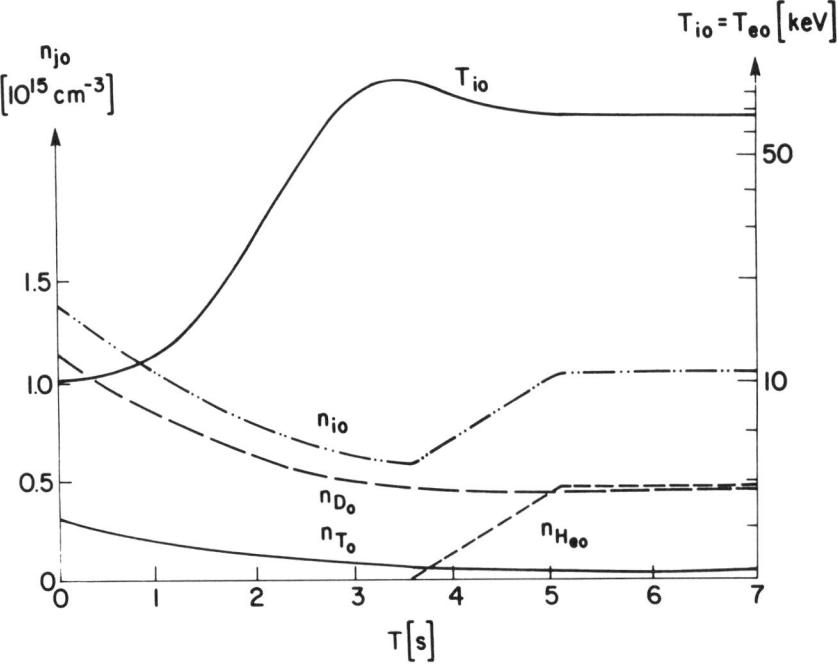

Source: S. Atzeni and B. Coppi, "Ignition Experiments for Neutronless Fusion Reactions," MIT Research Laboratory of Electronics Report PRR 80/11 (Cambridge, Mass., 1980), with minor changes.

Figure 15-2. Heating Strategy for the Proposed Candor Experiments in Order to Test D-He³ Ignition Conditions as Described in Section 3

due to Coulomb scattering only, are relaxed when considering the following effects:

1. Nuclear scattering collisions (knock-on events) increase the fraction of energy transferred to the background ions during the slowing down of the fast-charged fusion products in the background plasma.
2. The fuel particles promoted in energy by their interaction with the fusion-reaction products have a nonnegligible self-interaction probability (tail-tail events).
3. Fast fuel particles that are promoted in energy by a large energy-transfer nuclear-scattering event, or intermediate fusion-reaction products, can fuse with other fuel particles belonging to the Maxwellian part of their distribution (fast-fusion or propagating-reaction events).[3]

Comparison with Present-Day Experiments

We may now ask what degree of extrapolation from present-day experiments is required in order to demonstrate ignition for a D-He^3 reactor. If we take the rate of collisionality as one measure of extrapolation, we can argue that this is no more severe than that required to prove D-T ignition in low-particle-density, large-scale experiments such as those contemplated for the existing fusion program. In fact, the degree of collisionality, ν^{**}, that can be measured by the ratio of the torus major radius, R_0, to the particle mean free path is proportional to:

$$\nu^{**} \propto \frac{nR_0}{T^2}.$$

Thus, if we take $T_0 \sim 15$ keV, $n_0 \simeq 5 \times 10^{13}$cm^{-3}, and $R_0 \simeq 3$ m, for a low-density D-T igniting experiment, while $T_0 \simeq 65$ keV, $n_0 \simeq 2 \times 10^{15}$cm^{-3}, and $R_0 \simeq 1$ m, for the corresponding D-He^3 experiment, we can see that the ratio of the two relevant values of ν^{**} is not far from unity.

On the other hand, the highest values of $\beta_0 = 8\pi p_0/B_0^2$ (p_0 being the peak plasma pressure and B_0 the average vacuum magnetic field) obtained so far in toroidal experiments appear to be as high as 30 percent,[4] whereas the values required for the D-He^3 ignition experiment under consideration are around 40 percent. A higher degree of extrapolation would be needed if, instead, we refer to the type of experiments reported by Murakami.[5]

The set of minimal dimensions for the kind of experiment indicated as "Kathartor" by Coppi,[6] to test D-He^3 ignition, can be identified on the basis of the best-known forms of the electron and ion thermal conductivities that, when included in the appropriate transport codes, reproduce well the parameters of present-day experiments.[7] These dimensions are for a device that we shall refer to as Candor I: $R_0 = 105$ cm (major radius of the toroidal plasma column); and $a \times b = 40 \times 55$ cm^3 (semiaxes of the elliptical meridian cross-section), whereas the plasma parameters are characterized by $n_{eo} \simeq 1.5 \times 10^{15}$ cm^{-3} (peak electron density), $T_{eo} \simeq T_{io} \simeq$ 65 keV (peak temperature), $\tau_E \simeq 0.8$ sec, $B_0 \simeq 120$ kG, $<\beta> \simeq 13.5$ percent. The evaluation of $<\beta>$ has been made by assuming radial distributions of both the plasma density and temperature similar to those observed in existing experiments.

The power balance that has been evaluated for this experiment (see figure 15-3) is:

Total power produced by fusion reactions	100 MW
Power loss by transport associated with charged particles	53 MW
Bremsstrahlung emission	33 MW
Synchrotron emission:	10 MW
Neutron emission:	4 MW

Note that synchrotron radiation emission is only 10 percent of the total energy loss and, in contrast to the case of possible advanced-fuel reactors examined in the past, does not represent an insurmountable difficulty.

In order to gain a safety margin by about a factor of 1.7 over the electron-energy containment time that can be extrapolated from present-day experiments, we have also analyzed the feasibility of a larger experiment, (Candor II) with the following plasma column dimensions:

$$a \times b \times R_0 = 50 \times 75 \times 120 \text{ cm}^3.[8]$$

In this analysis we have assumed the same value plasma parameters indicated earlier for the smaller device, expect for the plasma current, which has been raised to 8 MA.

We also note that RF systems with frequencies around 130 MHz, delivering about 10 megawatts of power to the plasma, as would be required for the Candor I experiment, are well within the present state of the art.

We have carried out the structural analysis necessary to verify that the high-field magnets required to produce the 120-kG toroidal field and the 6-MA current mentioned earlier can be achieved with metallic conductors, such as aluminum, cooled to cryogenic temperatures. The structural com-

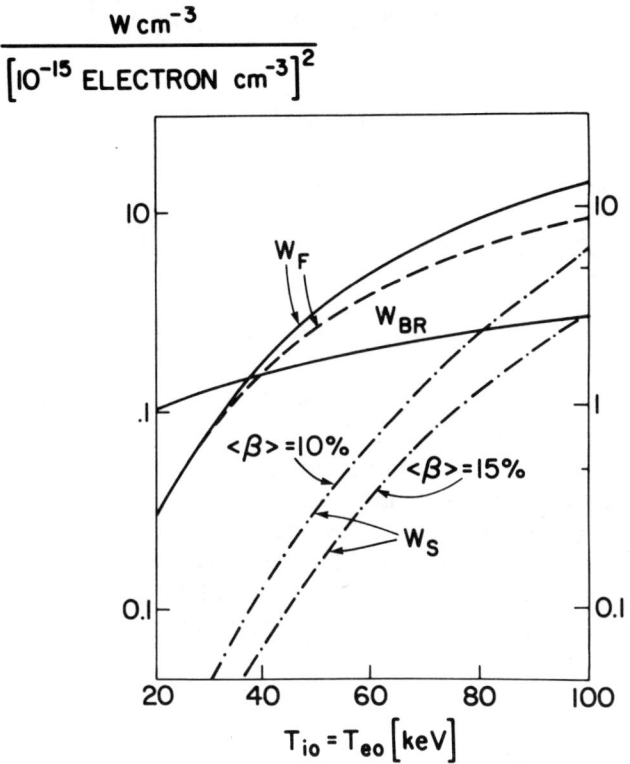

Source: S. Atzeni and B. Coppi, "Ignition Experiments for Neutronless Fusion Reactions," MIT Research Laboratory of Electronics Report PRR 80/11 (Cambridge, Mass., 1980), with minor changes.

Note: The dotted curve $\widehat{W}_F$ corresponds to evaluating the relevant reactivity by assuming that the fuel particles have a strictly Maxwellian distribution.

Figure 15-3. Representative Values of the Power Densities Produced by Fusion Reactions, ($\widehat{W}_F$), Lost by Bremsstrahlung, ($\widehat{W}_{BR}$), and by Synchrotron Emission, ($\widehat{W}_S$) in the Reference Candor Ignition Experiments as Functions of the Peak Plasma Temperature

ponents of these magnets, which carry the mechanical stresses, can be designed using the same criteria developed for the magnets used in compact D-T ignition experiments.[9]

A reaction cycle that can be realized with a device having nearly the same parameters as the candor experiments is the so-called catalyzed D-D reaction. This is described as follows:

$$D + D \begin{cases} \longrightarrow T\ (1.01\ \text{MeV}) + p\ (3.03\ \text{MeV}) \\ \\ \longrightarrow He^3\ (0.82\ \text{MeV}) + n\ (2.45\ \text{MeV}) \end{cases}$$

with

$$D \rightarrow He^4\ (3.52\ \text{MeV}) + n\ (14.6\ \text{MeV})$$

$$D \rightarrow He^4\ (3.67\ \text{MeV}) + p\ (14.67\ \text{MeV}),$$

and involves the "burning" of the charged products of the two branches of the D-D reaction, while the α-particles and the energetic protons emitted by the D-He^3 reaction are contained in order to compensate for electron thermal-energy loss by thermal conductivity, radiation processes, and so forth, as in the case of the D-He^3 reactor. A D-D reactor in which a considerable fraction of the produced He^3 can be extracted can be identified as a partially catalyzed D-reactor. The process of catalysis in these cases indicates that of reinjecting all or part of the He^3 that may be lost through diffusion from the reacting plasma column.

Power-Producing Systems

One objection to the possible practical use of D-He^3 reactors is that He^3 is very rare in nature. The problem of He^3 supply has been studied; and it has been suggested that the fuel for a D-He^3 reactor can be provided by an off-site, partially catalyzed deuterium reactor of equal power.[10] If a lithium blanket is added to the latter type of reactor, and decay tanks for the storage of tritium are included in the system, then an off-site reactor can fuel four D-He^3 reactors of equal power.[11] When compared to the D-T reactor, the D-D catalyzed reactor has the following advantages:

1. Its fuel is available in nature.
2. It does not need a lithium blanket to sustain itself.
3. The fraction of energy released in the form of neutrons amounts to about 40 percent per D-D reaction versus 80 percent in the case of a D-T reaction.
4. The tritium inventory is considerably lower.

In this connection, note that the fraction of energy produced in the form of neutrons in a D-He^3 reactor has been estimated at between 1 and 4 percent. The tritium throughput in a D-D reactor is smaller by a factor of 15 than that in a D-T reactor of equal power,[12] whereas the reduction factor is about 1,000 for a D-He^3 reactor. Another significant advantage is that the shielding-blanket requirements are considerably reduced for D-He^3 reactors compared with D-T reactors, making the characteristics of the toroidal magnet (containing both the plasma chamber and the blanket) compatible with the limits of present-day technology.[13] It has also been pointed out that confinement configurations in which $<\beta>$ is relatively high while the confining magnetic field is also high may not be exploitable for D-T reactors because of wall-loading problems associated with the high reactivity level of the relevant D-T plasmas.[14]

Thus from our perspective experimental D-T reactors appear to provide the most immediate means of acquiring the still-needed knowledge of the physics of thermonuclear plasmas. However, on a long-term basis, if the feasibility of candid reactors can be shown, D-T reactors may not be the most desirable systems for producing power.

Notes

1. H.W. Kendall and S.J. Nadis, eds., *Energy Strategies: Toward a Solar Future* (Cambridge, Mass.: Ballinger, 1980).

2. B. Coppi, J. Crew, and J. Ramos, "Search for the Beta Limit," Massachusetts Institute of Technology, Research Laboratory of Electronics Report PRR 80/19 (Cambridge, Mass., 1980). To be published in *Comments on Plasma Physics and Controlled Fusion Research*.

3. R.W. Conn and G.W. Shun, Paper V-5, *Proceedings of the Eighth International Conference on Plasma Physics and Controlled Nuclear Fusion Research* (Brussels, 1980). To be published by the International Atomic Energy Agency (Vienna).

4. C.K. Chu et al., Paper L-4-2, *Proceedings of the Eighth International Conference on Plasma Physics and Controlled Nuclear Fusion Research* (Brussels, 1980).

5. M. Murakami and ISX-B Group, Paper N-1, *Proceedings of the Eighth International Conference on Plasma Physics and Controlled Nuclear Fusion Research* (Brussels, 1980).

6. B. Coppi, "Transport Processes in High Current and Particle Density Plasmas," Massachusetts Institute of Technology, Research Laboratory of Electronics Report PRR 80/11, published in the *Proceedings of the 1980 International Conference on Plasma Physics* (Nagoya, Japan: Nagoya University, 1980), vol. II, p. 33.

7. S. Atzeni and B. Coppi, "Ignition Experiments for Neutronless Fusion Reactions," MIT Research Laboratory of Electronics Report PRR

80/11 (Cambridge, Mass., 1980). To be published in *Comments on Plasma Physics and Controlled Fusion Research.*

8. Ibid.

9. B. Coppi and A. Taroni, "Burn Phase and Structural Problems of Compact Fusion Reactor Experiments," Massachusetts Institute of Technology, Research Laboratory of Electronics Report PRR 79/22 (Cambridge, Massachusetts, 1979). To be published in B. Coppi et al., eds., *Physics of Plasmas Close to Thermonuclear Conditions* (Oxford: Pergamon Press, forthcoming).

10. G. Miley et al., Paper V-5, *Proceedings of the Eighth International Conference on Plasma Physics and Controlled Nuclear Fusion Research* (Brussels, 1980).

11. Ibid.

12. C.C. Baker, et al., Paper 2E3-4, *Proceedings of the Eighth International Conference on Plasma Physics and Controlled Nuclear Fusion Research* (Brussels, 1980).

13. Atzeni and Coppi, "Ignition Experiments."

14. J.R. Roth and H.C. Roland, personal communication.

16 Inertial Fusion and Energy Production

John F. Holzrichter

Introduction

Inertial confinement fusion (ICF) is a technology for releasing nuclear energy from the fusion of light nuclei. For energy production the most reactive hydrogen isotopes, deuterium (D) and tritium (T) are commonly considered. The energy application requires the compression of a few milligrams of a DT mixture to great density, ~1,000 times its liquid-state density, and to a high temperature, nearly 100,000,000°K. Under these conditions, efficient nuclear-fusion reactions occur, which can result in over 30-percent burn-up of the fusion fuel.[1] The high density and temperature can be achieved by focusing very powerful laser or ion beams onto the target. The resultant ablation of the outer layers of the target compresses the fuel in the target (see figure 16-1), DT ignition occurs, and burn-up of the fuel results as the thermonuclear burn wave propagates outward.

The DT-fuel burn-up occurs in about 100 picoseconds (100 millionths of a millionth of a second). On this short time scale inertial forces are sufficiently strong to prevent target disassembly before fuel burn-up occurs. The energy released by the DT fusion is projected to be several hundred times greater than the energy delivered by the driver.

The majority of the energy release is carried away from the target by 14.1-MeV neutrons (70 percent) but also by X-rays (~10 percent) and target debris (~20 percent). The α particles released in the reaction are stopped in the target. A typical reactor-size target might release 600 million joules (MJ) of energy, equivalent in energy to about 0.15 ton of exploding TNT. The impulse transmitted to the chamber walls by the fusion-target release is very much smaller than the impulse of exploding 0.15 ton of TNT because the fusion-target mass is so much smaller. The tremendous energy content of DT fuel is exemplified by comparing the 600 million joules of energy from an approximately 2-mm-radius fuel sphere in the aforementioned target with the previously described 300 pounds of TNT or, more practically, with the 6,000 MJ of energy in one 42-gallon barrel of oil. Although the oil barrel contains about ~10 times more energy than the fusion target, the barrel contains 3 million times more fluid. The cost of the deuterium fuel is insignificant (<< 1¢) compared with $30/barrel for oil. The tritium would be

---

Research performed under the auspices of the U.S. Department of Energy, Contract no. W-7405-Eng-48.

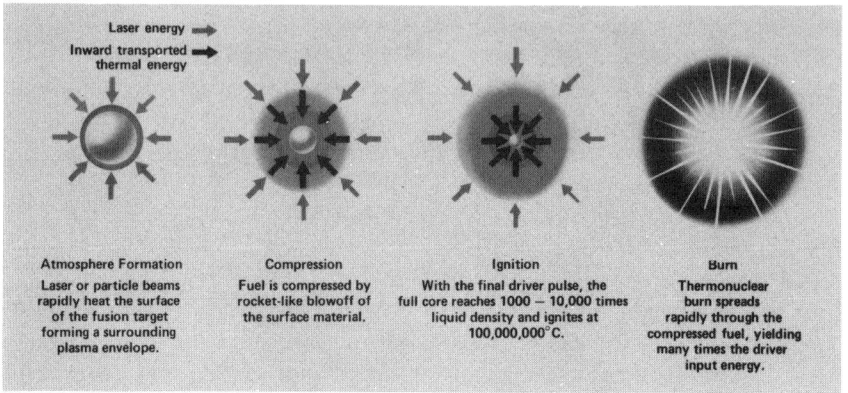

Figure 16-1. Inertial-Confinement Fusion Concept

costly except that an ICF reactor can generate enough tritium through the interaction of fusion-produced neutrons with lithium in the reactor wall. The target cost, which is governed by the capital cost of the target production line, is discussed later.

The energy release from the fusion target can be used for nuclear-weapons-effects simulation and for commercial power generation. The military aplications, which also include materials-response studies, physics understanding, and simulation, are the least demanding on the ICF technologies. They require a driver of sufficient energy delivery capability (~3 MJ) to provide a target yield of serveral hundred megajoules, but without the added driver requirements of high efficiency or high repetition rate, or the development of an economic commercial reactor technology. The laser-target interaction studies, the target design work, and the laser-technology developments that make up the majority of the work performed by the U.S. ICF community over the past ten years are virtually independent of the final ICF application. The demonstration of target ignition and the diagnostics of the fuel conditions with the Nova laser (now under construction) will narrow the uncertainty in target gain versus laser energy to an acceptable level (see figure 16-2). These experiments will provide enough information to plan construction of a driver at the several-megajoule level for a demonstration of high-gain target output. This is the scientific-feasibility experiment.

The commercial fusion-reactor application holds great promise because it is one of three inexhaustible power options (the others are solar power and the fission breeder cycle). Power generation based on fusion offers the potential for the inexhaustible fuel, deuterium. Deuterium is present in the molecule HDO in the world's water supply in the ratio of two HDO to 6,500 to H_2O molecules. Thus for all practical purposes it is inexhaustible.

Inertial Fusion and Energy Production

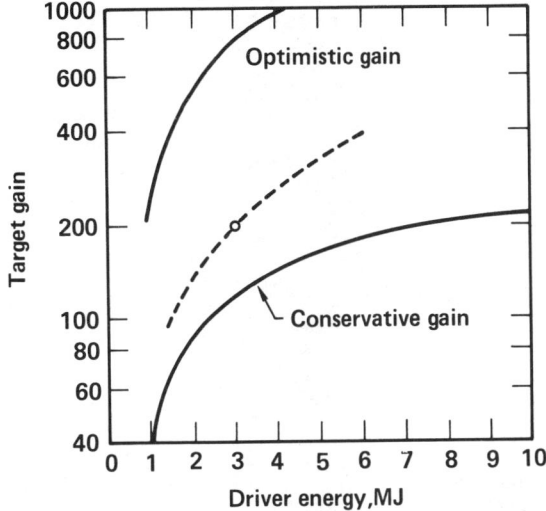

Note: Because the costs for the first multimegajoule driver will be large, it is important to minimize the uncertainty in its size.

Figure 16-2. Calculated Target Gain as a Function of Driver Energy

Tritium, which is mixed with deuterium to fuel the targets, is generated (or bred) from lithium in the reactor system to provide the complete fuel cycle.

The attractiveness of the ICF reactor concept is that the reaction chamber is separated from the complex fusion-ignition machinery, which includes the laser driver or ion-beam driver (see figure 16-3). In addition, the reactor system can be designed to minimize neutron-induced radioactivity, to minimize the tritium inventory, and to operate at a very moderate vacuum level.

Present Status

The last ten years of ICF technology development have been very fruitful. Target physics experiments and construction of demonstration facilities evolved starting with the proof of fusion temperatures in low-compression targets in 1974.[2] Shortly thereafter, proof of thermonuclear-reaction experiments showed that the targets were working according to theory by measuring the energy distribution of the α-particle and neutron emission.[3] The high-density experimental campaign demonstrated fuel compression to greater than 100 times liquid DT density in low-temperature targets.[4] This density is within a factor of 10 required for high-gain targets. Recent experiments are confirming theoretical predictions and are showing more effi-

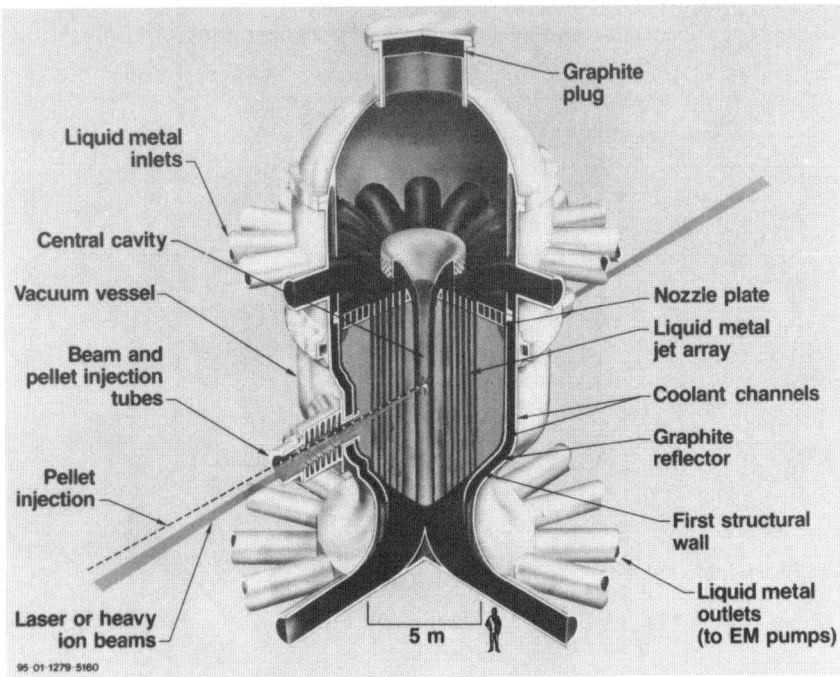

Note: The Livermore HYLIFE reactor concept is a liquid-metal-walled reactor chamber that absorbs the intense neutron and X-ray energy released from the target and thermalizes it for delivery to a steam-turbine system.

Figure 16-3. HYLIFE Reactor Chamber

cient target coupling with shorter wavelength laser drivers.[5] Figure 16-4 summarizes ICF target performance and compares it with magnetic-confinement fusion-machine performance.

Driver technology has developed along two different lines. The Nd:glass laser has been developed to a mature technology as the main experimental tool for target-interaction and fuel-compression studies.[6] Its advantages for this task have been its relatively advanced state of development in the early 1970s when an experimental driver with multiterawatt, multikilojoule capabilities was required. Its ability to produce target intensities of 10^{14} W/cm^2; its attractive wavelength (1 μ); and the potential to use its harmonic wavelengths (0.5 μ, 0.35 μ, 0.26 μ) for plasma heating were all desirable. The glass-laser-technology development has proceeded rapidly with a succession of systems constructed or being constructed at Livermore—Argus at 2 kJ, Shiva at 15kJ, and Nova at 250 kJ—and elsewhere in the United States at the University of Rochester, KMS Fusion, and NRL. Other glass-laser facilities are in England at AWRE and Rutherford Labs, in France at Ecole

Inertial Fusion and Energy Production

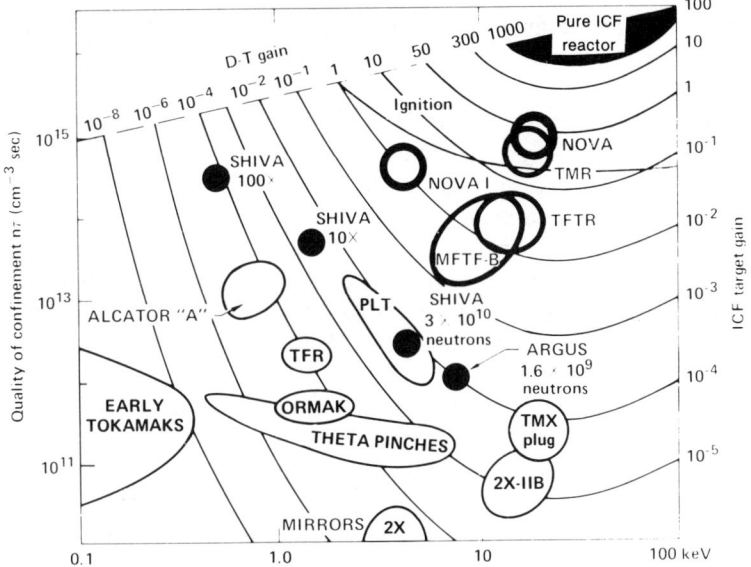

Note: Progress and projections of performance for ICF and MFE experimental facilities in achieving the conditions necessary for a fusion reactor. The plasma conditions achieved in the Shiva targets are the same as in PLT, the Princeton Large Torus (Tokamak). For reactor systems ICF must reach a higher target gain than MFE systems because of the need to run the driver.

Figure 16-4. Thermonuclear Conditions Achieved in Fusion Experiments

Polytechnique and CEA Limeil, in Japan at Osaka University, and in the USSR at the Lebedev Laboratory. The second line of driver-technology development emphasizes efficiency and repetition rate. To this end, CO_2 laser technology is being pursued at Los Alamos Scientific National Laboratory (LASNL) with the execution of the Helios laser, which provided 6 kJ of $10 = \mu =$ wavelength laser energy on target and the present construction of the Antares laser for 40-kJ operation.

An important class of lasers that are being developed and considered for fusion applications are the rare-gas halide systems—KrF, XeF, XeCl. These lasers have a potential of > 5-percent system efficiency and high-repetition-rate capability; and their wavelengths are attractive from the point of view of providing efficient, classical coupling to the target.

A most important realization in the driver area was that high-energy particle-accelerator technology could be extended to the high-power, high-energy, pulse regime.[7] This mature technology developed by the high-energy physics community has the potential to provide a > 10-percent, high-repetition-rate driver for the commercial reactor application. Along with these technical innovations in new driver technologies, reasonably consistent cost/performance analysis is now occurring, which permits rational comparisons to be made between competing and expensive driver technologies.

Finally, a reactor system, the HYLIFE design described earlier, has been invented and is being developed.[8] This approach promises to provide a reactor system that has a thirty-year life, is capable of handling the target impulse, and provides complete decoupling between the driver and the reaction chamber. With a target-gain times driver-efficiency product of 20 (200 gain × 0.10 driver efficiency), an ICF driver can be compared with the liquid-metal fast breeder reactor (LMFBR) concept in terms of electricity costs, if the ICF-reactor costs equal the LMFBR-reactor costs and the ICF-driver and pellet-factory costs add less than 30 percent of the LMFBR cost to the ICF system. The 30-percent capital-cost differential is due to the costs of the nuclear fuel, safety, and reprocessing.[9]

ICF technology has undergone several review processes, with the most recent reviews by the Ad Hoc Experts Committee on Fusion led by Dr. John S. Foster of TRW Incorporated. The committee's report emphasizes the need to aggressively pursue ICF research. They see the ICF approach as a successful path to energy production without scientific or technological problems that would prevent demonstration of feasibility. Admittedly, it could turn out to be otherwise, but there is an urgency to knowing whether or not ICF is a real energy option.

Notes

1. J.H. Nuckolls, L.L. Wood, A.R. Thiessen, and G.B. Zimmerman, "Laser Compression of Matter to Super-High Densities: Thermonuclear (CTR) Applications," *Nature* 239, no. 5368 (1972):139-192; K.A. Brueckner and S. Jorna, "Laser Driven Fusion," *Review of Modern Physics* 46, no. 325 (1974); and J.L. Emmett, J.H. Nuckolls, and L.L. Wood, "Fusion Power by Laser Implosion," *Scientific American* (June 1974):24-37.

2. G. Charatis, J. Downard, R. Goforth, B. Guscat, T. Henderson, S. Hildum, R. Johnson, K. Moncur, T. Leonard, F. Mayer, S. Segal, L. Siebert, D. Solomon, and C. Thomas, "Experimental Study of Laser Driven Compression of Spherical Glass Shells," *Plasma Physics and Controlled Nuclear Fusion Research* 2 (1974):317 (Vienna: IAEA, 1975). (Proceedings of the Fifth IAEA Plasma Fusion Conference in Tokyo, 1974.) J.F. Holzrichter, H.G. Ahlstrom, R.D. Speck, E.K. Storm, J.E. Swain, L.W. Coleman, C.D. Henricks, H.N. Kornblum, F. Seward, V.W. Slivinsky, Y.L. Pan, G.B. Zimmerman, and J.H. Nuckolls, "Implosion Experiments with an Asymmetrically Irradiated Laser Fusion Target," *Plasma*

Physics 18 (1976):675-680; K.R. Manes, H.G. Ahlstrom, R.A. Haas, and J.F. Holzrichter, "Light-Plasma Interaction Studies with High-Power Glass Lasers," *Journal of the Optical Society of America* 67 (1977):717.

3. V.W. Slivinsky, H.G. Ahlstrom, K.G. Tirsell, J. Larson, S. Glaros, G. Zimmerman, and H. Shay, "Measurement of the Ion Temperature in Laser Driver Fusion," *Physical Review Letters* 35 (1975):1083; R.A. Lerche, L.W. Coleman, J.W. Houghton, D.R. Speck, and E.K. Storm, "DT Ion Temperatures Determined from Neutron Energy Spectrum of Laser Fusion Targets," *Applied Physics Letters* 31 (1977):10.

4. H.G. Ahlstrom, "Fusion Experiments at LLL," UCID 18707 (June 1980).

5. Ibid.

6. J.F. Holzrichter, "High Power Pulsed Lasers," UCRL-52868 (April 1980). Lectures given at the Twentieth Scottish Summer School in Physics on Laser Plasma Interaction, also published in its proceedings.

7. A.W. Mashke, Brookhaven National Laboratory Report, BNL 19008 (1974). Also see reports from workshops on heavy-ion fusion: Lawrence Berkeley Laboratory Report LBL 5543 (1976); Brookhaven National Laboratory Report BNL 50769 (1977); and Argonne National Laboratory Report ANL 79-41 (1979).

8. M.J. Monsler, J.A. Maniscalco, J.A. Blink, J. Hovingh, W.R. Meier, and P.E. Walker, "Electric Power from Laser Fusion: The HYLIFE Concept," *Proceedings of the Thirteenth Intersociety Energy Conversion Engineering Conference*, San Diego, Calif., 1978, p. 2042. M. Monsler and W. Meier, "A Conceptual Design Strategy for Liquid-Metal-Wall Inertial Fusion Reactors," accepted for publication by Nuclear Engineering and Design. Also UCRL-84881.

9. C.L. Rudasill, "Comparing Coal and Nuclear Generating Costs," *EPRI Journal* (October 1977).

17 Fusion and U.S. Energy Policy

N. Douglas Pewitt

This chapter will address primarily the nearer-term realities of the U.S. fusion program. This should provide relevant background material for the rest of the book. Certainly these near-term prospects are relevant to the future of fusion as an energy source, not only because the U.S. fusion program is the largest single national program, but also because we are in the final stage of cementing a national commitment to the development of magnetic fusion.

This chapter will address the current status of the fusion program and the considerations that have made fusion-engineering development the appropriate next step toward the goal of achieving the highest potential for fusion at the earliest practical date. I will address these issues, however, in the overall context of the Department of Energy's responsibilities and the inescapable competition for funds and other resources. Despite the promise that fusion holds and the progress that is being made, it is but one option among many that must be addressed both by the department and by the broader technical community. Furthermore, research and development is only one of the avenues through which the U.S. government can help meet the nation's energy needs.

In the seven years since the 1973 oil embargo, the United States has been buffeted by a series of shocks that have forced us to recognize more clearly the complexities of the energy situation, both domestically and worldwide. Broadly speaking, the energy problem has three basic facets: (1) high and increasing oil prices, (2) vulnerability to supply disruption, and (3) uncertainty about future energy prices. The United States has been moving forward on several fronts to deal with these issues. Increased prices, conversion to other energy sources, favorable weather, and improved efficiency of energy use have led to almost a 20-percent reduction in oil imports as of late-1980. But there is no reason to believe that this can afford us more than temporary relief. The United States remains vulnerable to further economic and political shocks from shifts in oil supply.

Consequently, there are four broad responsibilities that the Department of Energy (DOE) must address:

1. short-term contingency planning;
2. oil-import reduction strategies;
3. increase and diversification of foreign energy supplies;

4. funding of R&D and other activities that reduce the lead times and uncertainty associated with alternative energy sources.

Fusion is just one of many approaches being addressed in this last area. In advising the secretary of energy on R&D matters, we must consider the problem of priorities, particularly for the budget, in meeting the broad responsibilities listed here. This involves a host of different trade-offs in which apples and oranges must necessarily be compared—regulatory versus R&D, conservation versus supply, near term versus long term, dispersed versus centralized, and so forth. Obviously, there is no unique or closed solution to this problem, and simplifying assumptions must be made. For example, the department must carry out its legislated regulatory responsibilities, address contingency planning adequately, maintain its commercial business responsibilities, get its near-term house in order as fast as is feasible, focus on the immediate problem of transportation liquids, ensure that environmental issues are being addressed properly, and so forth.

The department has some eighty separate major budget categories relating to its energy responsibilities. These can be grouped into eighteen functional clusters—regulatory activities, emergency planning, liquids-near term, oil backout-near term, and so on—which can then be rank ordered on the basis of reasonable criteria that flow from the assumptions. Within each cluster the individual programs can be further rank ordered, again on the basis of reasonable criteria, such as technical status, deployment, urgency, leverage of the government, cost, and so forth. Such analysis is not a precise tool, but it does provide a conceptual framework for thinking about priorities.

However, despite its apparent promise, within this grouping, magnetic fusion was ranked number 70 out of 80. How can we reconcile such analysis with the intuition of many observers that fusion is a promise of the future? To deal with this apparent contradiction, let me discuss in more detail the context in which fusion must compete.

Fusion is primarily thought of as a source of long-term electricity supply, although there are other potential routes through which fusion might reach its highest potential. For fusion to be successful, it must prove to be environmentally and economically preferable to other alternative energy systems. The competition is stiff. For example:

> Electricity from coal is available today, and significant investments are being made in coal cleanup and stack-gas cleaning technologies that would enhance the usefulness and acceptability of conventional steam-fired plants. New boiler types such as atmospheric and pressurized/fluidized beds are being developed and tested. Fuel cells are being developed that might use coal-derived liquids. Many utilities are exploring

cogeneration and gasified coal-combined cycle systems that have the promise of higher thermal efficiencies.

Nuclear generation is a commercial reality with the conventional light-water-reactor (LWR) technology. The DOE is supporting work to improve these systems. Initiatives to deal with radioactive waste and the various steps being taken subsequent to Three Mile Island may begin to restore public confidence in nuclear power. R&D is continuing on fast breeder reactors, though at a less aggressive pace than in the past because of current views about the availability of uranium for the LWR fuel cycle.

Solar and geothermal development is continuing and is bound to play some role in future electricity markets.

Each of the systems mentioned has problems, economic or environmental. Coal-fired systems produce CO_2 that may have undesirable effects on the world's climate. Scrubbing and disposing of the CO_2 would roughly double the price of electricity from coal. Scrubbing oxides of sulfur already adds significantly to electricity costs. For fusion to be economically and environmentally preferable, it would thus need to produce electricity, as an upper bound, at a price no more than twice the projected price of current coal-fired systems.

The nuclear fuel cycle has environmental and safety problems also. Nevertheless, for fusion to be preferable, it will need to be cost competitive and superior in environmental and safety features to the fission systems that are available at the time of introduction. The potential availability of a wide range of alternative sources of electricity, each of which is moving on a faster track than fusion with respect to initial introduction, makes it highly important to determine whether fusion truly can be a competitive option for the production of electricity—that is, whether there are insurmountable barriers of engineering performance and design that will keep fusion from being cost effective.

These economic issues are emphasized because the ultimate decisions about the commercialization of energy systems will be economic and investment decisions. The end-use market is also important. The electric-utility industry is currently in sad financial shape and, in addition to the economic pressures on the industry, there is a significant timing issue connected with anticipated growth.

Utility planners, who must work on at least a twenty-year time scale, originally made construction commitments based on historical rapid growth. But escalating prices, coupled with vigorous conservation measures, have sharply decreased demand growth. Hence some utilities are seeing zero growth, or even negative growth. Because of construction decisions made twenty years ago, fuel prices that they now cannot afford, and lags in establishing rates that reflect true costs, many power companies are now in serious trouble. There is a large backlog of currently planned or committed capacity to be worked off. At the same time, some regions of the country, such as the Northwest, and some particular utilities are facing severe shortages of capacity and will have serious near-term problems in meeting their responsibilities to provide reliable, economic service.

It is thus very difficult to say when the demand for electricity will be sufficient to justify the introduction of a radically new technology such as fusion. Moreover, utility needs are shaped by other issues that go beyond technical availability and financial capability. In integrating new technology with their systems, utilities must be concerned about plant size, complexity, and operational characteristics. Consequently, fusion systems will need to be carefully matched to the needs of utility systems.

In light of all the foregoing considerations, why should the United States now proceed with fusion-engineering development or even continue at the present funding level? The answer hinges on the long-term, high-cost character of fusion, the underlying logic of technological development, and the promise fusion holds for future generations. In order to sustain a major commitment in this area, it is necessary that the nation have an overall policy supported by a broad consensus. A fundamental aspect of such a policy must be to reduce uncertainty—that is, to determine as early as practically possible whether fusion is indeed likely to be a viable option in terms of both cost and performance.

If for some reason the promise of fusion turns out to be dim, then the very high cost phase of commercial-scale demonstration can be avoided, and the nation will know early that it will have to rely on other sources. If, on the other hand, the practical promise of fusion turns out to be bright, then the whole perspective of the world's long-range energy future would be changed. This implies that, to limit cost and reduce uncertainty, we should move ahead with engineering development as soon as it makes technical sense to do so. Until such a step is taken, fusion will remain the most uncertain of the competing energy systems, with the least known about its feasibility.

Up to 1978 the basic policy on fusion was a linear approach leading to power generation. Under this concept the program would proceed from proof-of-principle experiments to major scaling experiments and then to energy-producing experimental reactors leading ultimately to commercial power-producing systems. This implied a lengthy and expensive "horse race" among competing concepts.

In 1978, after a careful review by a committee chaired by John Foster, a formal policy was developed that embodied a more cautious approach, establishing an overall goal to determine the highest potential employment

Fusion and U.S. Energy Policy

of fusion energy through a strategy of pursuing several physics approaches and parallel technology development while continuing to preserve and develop the technical base and delaying a commitment to a major new system. The near-term objectives could be summarized as follows:

1. Demonstrate scientific feasibility.
2. Establish a sound engineering base.
3. Maintain a sound scientific base.
4. Encourage alternate concepts.

During the past two years the U.S. magnetic-fusion program has made rapid technical progress toward these goals; and confidence has been greatly increased that major physics experiments now under construction will be successful. In February 1980, in light of this rapid progress, Ed Frieman requested the Energy Research Advisory Board of the DOE to review two aspects of the magnetic-fusion program: first, the choice of the next major steps toward economic fusion power and, second, the overall strategy of the program. The conclusions of the panel, chaired by Sol Buchsbaum, endorse an acceleration of the magnetic-fusion program into parallel phases of engineering and physics development. The basic conclusions include:

1. Recent progress in plasma-confinement research is impressive.
2. Demonstration of scientific feasibility is near.
3. A burning-plasma device can be built.
4. An optimum fusion reactor cannot yet be specified.
5. The safety, economics, and environmental acceptability of fusion have yet to be demonstrated.
6. The United States is now ready to embark on the next step toward economic fusion power: exploration of the engineering feasibility of fusion.

Implementation of these recommendations requires adding a parallel component of engineering development to the existing effort of scientific and technological development. The engineering program will center around a tokamak-based Fusion Engineering Device (FED) that would provide a burning, perhaps even an ignited, plasma; provide a focus for developing and testing reactor-relevant technology and components; explore problems of operator and public safety; and serve as a focus for a broadly based program of engineering experimentation and analysis. These recommendations imply significant changes in the magnetic-fusion program, including a doubling of cost in real terms over the next five years and a commitment to a billion-dollar expenditure for the Fusion Engineering Device.

The approach put forward by the Buchsbaum panel greatly increases the likelihood that the role of fusion in the energy future can be assessed with some certainty in about ten years time. That assessment is crucial to the overall development of national energy policy. Energy systems have enormous inertia, and the time required for widespread commercial introduction of the technology will be somewhere from ten to thirty years after information from an FED would be available. For fusion to make a significant energy contribution by 2020, energy policymakers, including the private sector, must have a clear understanding of its promise in the 1990s.

Overall fusion policy has been modified in order to implement these recommendations. This step has been approved by the administration, and legislation formalizing this policy has to be enacted. The overall goal of the program remains unchanged: to develop the highest potential of fusion that is both environmentally sound and economically competitive. However, with the first major goal of the program, demonstration of scientific feasibility, now thought to be assured, we are now preparing plans to move fully into an engineering-development phase focused around a Center for Fusion Engineering (CFE), where an FED will be designed and constructed along the lines recommended by the Energy Research Advisory Board.

The strong language in the original policy concerning the importance of the breadth of the program and of the strength of the scientific base remains fully appropriate. Consequently, although a generic reactor-engineering base is being developed through the CFE, research in alternative confinement schemes and advanced tokamak physics will continue; and theoretical and experimental physics research essential to further advances will be supported and protected. Adoption of this policy provides the basis for achieving the second major goal of the fusion program—namely, the determination of the engineering practicality of fusion—and provides a concrete, near-term focus for the program.

The elements of a national consensus to move forward with this new fusion policy appear to be in place. Nevertheless, it should be remembered that this commitment must ultimately be maintained over decades; that the competition for resources will be keen; and that, in pressing forward in areas that go well beyond previous practical experience, there are bound to be unpleasant surprises. We will need to keep firmly in mind the true significance of fusion: that we are making investments now that can be crucial to the future not only of the United States, but also of the entire world community.

18 Status of U.S. Tokamak Effort

Paul J. Reardon

The work reported on in this chapter is almost fully supported by the U.S. Department of Energy's Office of Fusion Energy. Without their enthusiasm, guidance, and dedication, or their continuing support, the U.S. tokamak program would neither be as strong as it is nor be the most productive in the world. Similarly, the excellent work reported on here (and much that could not be reported on) is the result of the theoretical, experimental, technological, engineering, and management endeavors of an extremely hard-working team of dedicated individuals who are fully committed to the development of fusion power. I am particularly indebted to the management and staff of Princeton University Plasma Physics Laboratory (PPPL), Oak Ridge National Laboratory (ORNL), Massachusetts Institute of Technology (MIT), General Atomic Company (GA), Lawrence Berkeley Laboratory (LBL), and Argonne National Laboratory (ANL) for their assistance in the preparation of this chapter and in some cases for the text, figures, and photographs reproduced here from their published and unpublished reports. The list of reports and authors would be too long for this chapter, and any list of individual names would surely omit some key participants. Finally, I must gratefully acknowledge the assistance of the PPPL Public Information Office, Graphic Arts Group, and Photo Shop, and the Headquarters staff of the PPPL Technology Department.

Introduction

After a brief introduction to fusion power and plasma physics and a description of tokamak magnetic confinement, this chapter summarizes the status of the U.S. experimental program with major tokamak devices. It will then describe the TFTR, the next major step in the U.S. tokamak program, which constitutes the critical break-even fusion-power experiment and demonstrates its scientific feasibility. The characteristics of the Princeton Tokamak Fusion Test Reactor (TFTR) will be compared with other major tokamaks also under construction abroad and with the proposed Engineering Test Facility. Some observations about the Fusion Engineering Device (FED) and the proposed U.S. Center for Fusion Engineering (CFE) will conclude the chapter.

Advantages of Fusion

As a source of energy, sustained thermonuclear-fusion power would have many advantages:

1. The major fuel, deuterium, can be readily extracted from ordinary seawater, which is available to nearly all nations. The surface waters of the earth contain more than 10^{13} tons of deuterium—an essentially inexhaustible supply. The tritium also required to achieve fusion would be produced as a by-product of the fusion reaction from lithium, which is available from land deposits or from seawater, which contains thousands of years' supply. The worldwide availability of these materials would reduce international tensions caused by currently existing geographic imbalances in fuel supplies.
2. The amounts of deuterium and tritium in the fusion-reacting plasma will be so small that a large uncontrolled release of energy would be impossible. In the event of a malfunction, the plasma would strike the walls of its containment vessel and instantly cease producing nuclear reactions.
3. Since no fossil fuels are used, there will be no release of chemical-combustion products.
4. There will be no fission products formed to present a handling and disposal problem. Radioactivity will be produced by neutrons interacting with the reactor structure, but careful materials selection is expected to minimize the handling and ultimate disposal of activated materials.
5. There is no long-term storage problem with spent fuel materials as those gases not used up in the ongoing fusion reactions are returned to their on-site storage facilities for subsequent use.
6. Another significant advantage is that the materials and by-products of fusion are not suitable for use in the production of nuclear weapons.

The abundance of raw materials, their wide distribution, and the attractive environmental features of fusion are augmented by the expectation that fusion energy will also become an economically competitive source of electricity. The biggest potential problem is the release of radioactive tritium gas under severe fault or accident conditions, but it is believed that sufficient controls can be utilized to eliminate this potential hazard.

D-T Fusion Reaction

To produce net power, fusion reactions must, in a noncompression system, take place at high temperatures. The power-production process that can occur at the lowest temperature, and hence the most readily attainable fusion process on earth, is the combination of a deuterium nucleus with one of tritium. Three isotopes of hydrogen are known: hydrogen (H), deuterium (D), and tritium (T). The nuclei of all three isotopes contain one proton, which characterizes them as forms of the element hydrogen; in addition, the deuterium nucleus has one neutron, and the tritium nucleus has two. In each case the neutral atom has one electron outside the nucleus to balance the charge of the single proton (see figure 18-1).

The products are energetic helium-4 (^{4}He) the common isotope of helium (which is also called an alpha particle), and a more highly energetic free neutron (n). The helium nucleus carries one-fifth of the total energy released and the neutron the remaining four-fifths (see figure 18-2).

Conditions for Fusion

Since nuclei carry positive charges, they normally repel one another. The higher the temperature, the faster the atoms or nuclei move. When they collide at these high speeds, they overcome the force of repulsion of their positive charges, and the nuclei fuse. In such collisions, fusion energy is released.

The difficulty in producing fusion energy has been to develop a device that can heat the deuterium-tritium fuel to a sufficiently high temperature and then confine it for long enough that more energy is released by fusion than is used for heating the fuel in the first place.

Confinement

High as these temperatures are, they are attainable; the problem is how to confine the deuterium and tritium under such extreme conditions. Part of the solution to this problem lies in the fact that at the high temperatures re-

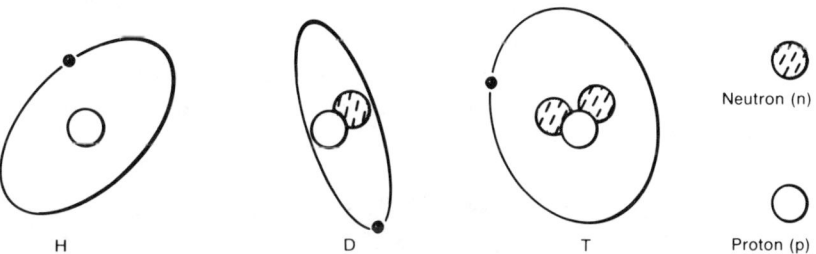

Figure 18-1. Isotopes of Hydrogen

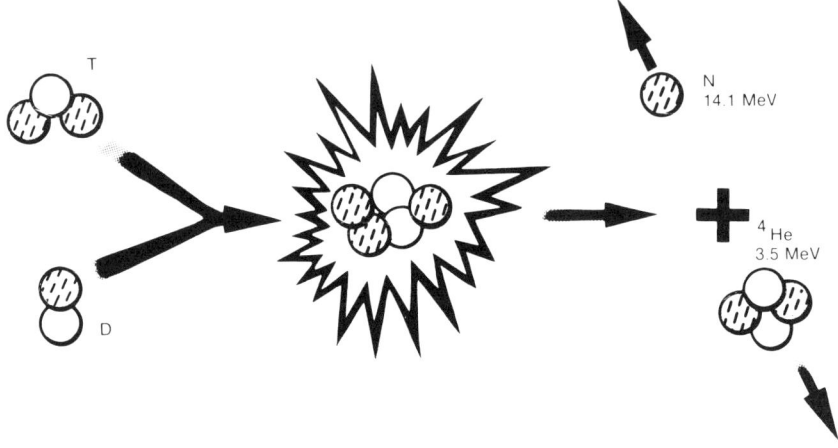

Figure 18-2. D-T Thermonuclear Reaction

quired, all the electrons of the light-hydrogen-atom isotopes become separated from the nuclei. This process of separation is called *ionization*, and the positively charged nuclei are referred to as *ions*. The hot gas containing negatively charged free electrons and positively charged ions is known as a *plasma*.

Because of the electric charges carried by electrons and ions, a plasma can, in principle, be confined by a magnetic field. In the absence of a magnetic field, the charged particles in a plasma move in straight lines and random directions. Since nothing restricts their motion, the charged particles can strike the walls of a containing vessel, thereby cooling the plasma and inhibiting fusion reactions. But in a magnetic field the particles are forced to follow spiral paths about the field lines. Consequently, the charged particles in the high-temperature plasma may be confined by the magnetic field and thus prevented from striking the vessel walls (see figure 18-3).

To produce substantial net power from fusion reactions, the following condition, known as the ignition condition, must be achieved; the product of the energy confinement time, τ (in seconds), and the plasma density, n particles (ions or electrons) per cubic centimeter, at a temperature of 100,000,000°C is such that:

$$n\tau \text{ must be greater than} \sim 3 \times 10^{14} \text{ cm}^3\text{-sec.}$$

Under these ignition conditions the alpha particles (helium nuclei) created in the thermonuclear fusion reactions are of sufficiently high temperature or

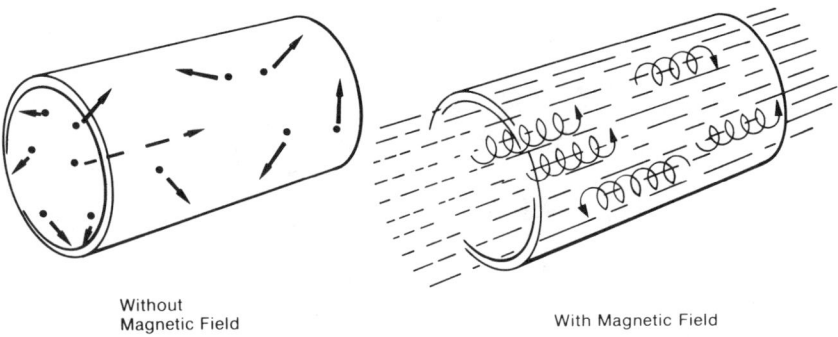

Figure 18-3. Plasma Confinement

energy (3.5 MeV) that they themselves, spiraling around the magnetic-field lines, heat the plasma, and the initial heating systems can be turned off as the alpha-particle heating keeps the plasma at the 100,000,000°C required for the fusion reactions to continue to take place.

Tokamak Plasma Confinement

Many different magnetic-confinement schemes have been studied during the past twenty-eight years. The one that is receiving the greatest attention in the U.S. magnetic-fusion energy program is the *tokamak concept*, which was first employed in the USSR.

The tokamak is a toroidal device that has a hollow, doughnut-shaped vessel through which magnetic fields twist, confining the plasma. Magnetic fields are produced by electric currents passing through wires or plasma. In the tokamak, there are two major components of the magnetic field. First, the toroidal magnetic-field component is generated by electric currents flowing in turns or rings around the torus. Second, the poloidal magnetic-field component is generated by a current flowing through the plasma (see figure 18-4).

An electric current in equilibrium-field coils outside the torus generates an auxiliary magnetic field that controls the position of the plasma in the torus and modifies the poloidal field. The equilibrium magnetic field is thus regarded as a component of the tokamak poloidal field.

By combining toroidal and poloidal magnetic fields, the tokamak confinement system achieves a higher level of plasma stability than has been

170 A Global View of Energy

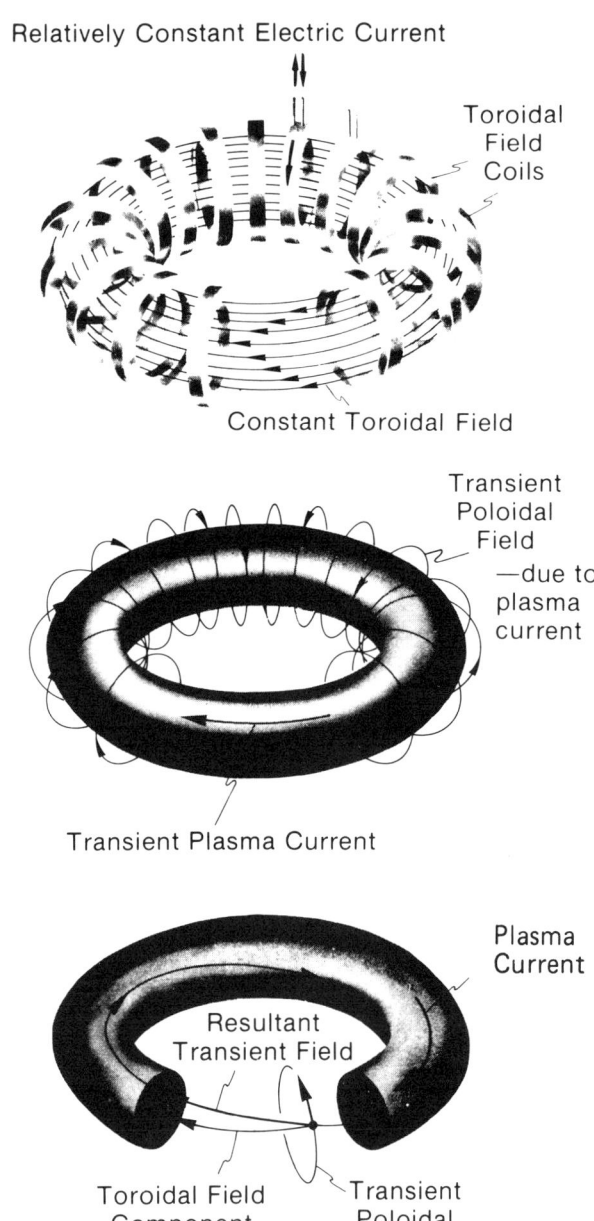

Figure 18-4. Tokamak Plasma Confinement

Status of U.S. Tokamak Effort

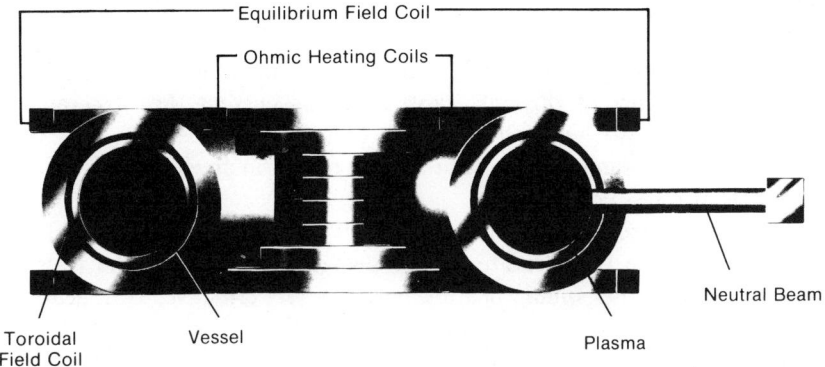

Figure 18-5. Tokamak System

realized in earlier magnetic confinement systems. It is this increased plasma stability that permits longer confinement times of higher-temperature plasmas (see figure 18-5).

Temperature

In order to release energy at a level of practical use for electricity production, the gaseous deuterium-tritium fuel must be heated to about 100,000,000°C. This temperature is more than six times hotter than the interior of the sun, which is estimated to be 15,000,000°C.

Plasma Heating

In an operating fusion reactor, part of the energy generated will serve to maintain the plasma temperature as fresh deuterium and tritium are introduced. However, in the start-up of a reactor, either initially or after a temporary shutdown, the plasma has to be heated to the required 100,000,000°C.

In current tokamak (and other) magnetic-fusion experiments, insufficient fusion energy is produced to maintain the plasma temperature. Consequently, the devices operate in short pulses, and the plasma must be heated by every pulse. (At present, Princeton has achieved tokamak plasma tempertures of 85,000,000°C.)

Ohmic Heating

Since the plasma is an electrical conductor, it is possible to heat the plasma by passing a current through it; in fact, the current that generates the poloidal field also heats the plasma. This is called *ohmic* (or *resistive*) *heating*; it is the same kind of heating that occurs in an electric light bulb or in an electric heater.

The heat generated depends on the resistance of the plasma and the current. But as the temperature of the heated plasma rises, the resistance decreases and other ohmic heating becomes less effective. It appears that the maximum plasma temperature attainable by ohmic heating in a tokamak is 20,000,000-30,000,000°C. To obtain still higher temperatures, additional heating methods must be used.

Neutral-Beam Injection

Neutral-beam injection involves the introduction of high-energy (neutral) atoms into the ohmically heated, magnetically confined plasma. The atoms are immediately ionized and are trapped by the magnetic field. The high-energy ions then transfer part of their energy to the plasma particles in repeated collisions, thus increasing the plasma temperature.

Magnetic Compression

A gas (or plasma) can be heated by sudden compression. In the same way, the temperature of a plasma is increased if it is compressed rapidly by increasing the confining magnetic field. In a tokamak system this compression is achieved simply by moving the plasma into a region of higher magnetic field (that is, radially inward). Since plasma compression brings the ions closer together, the process has an additional benefit of facilitating attainment of the required density for a fusion reactor.

Radiofrequency Heating

In radiofrequency heating, high-frequency waves are generated by oscillators outside the torus. If the waves have a particular frequency (or wavelength), their energy can be transferred to the charged particles in the plasma, which in turn collide with other plasma particles, thus increasing the temperature to the bulk plasma.

Status of U.S. Tokamak Effort

Alpha-Particle Heating

As mentioned earlier, when significant thermonuclear reactions take place, as will be the case with Princeton's Tokamak Fusion Test Reactor (TFTR), alpha-particle heating also occurs. When a plasma is ignited, the alpha particle self-heating features of the reacting fusion plasma are sufficient to sustain the thermonuclear burn (see figures 18-6 and 18-7).

Tokamak Experimental Program

Table 18-1 shows the three areas of physics which need to be attacked and resolved before one can establish a practical fusion reactor. We are now in the basic plasma physics period and the following seven figures and one table show the significant progress being made on existing tokamaks in all

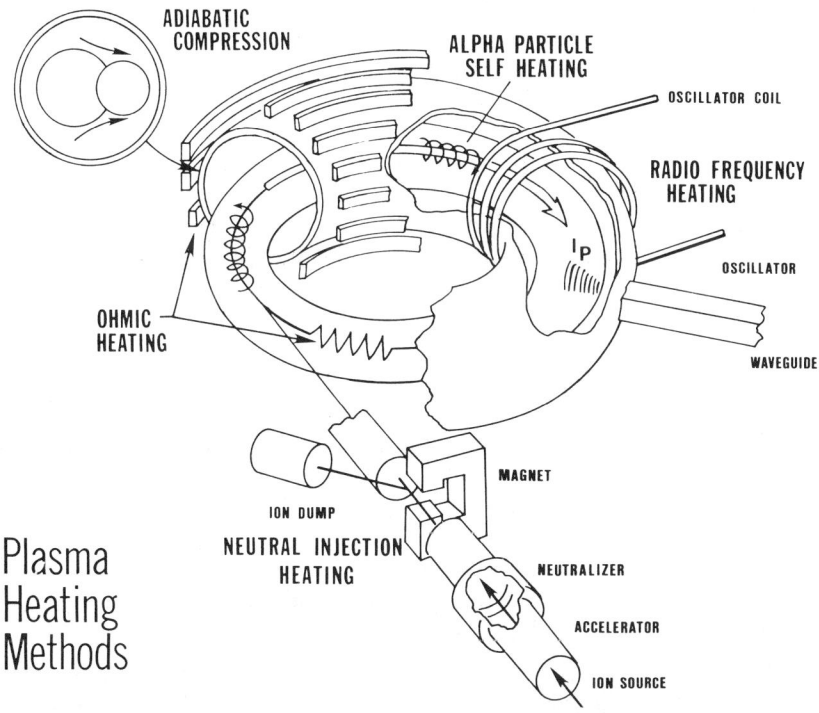

Figure 18-6. Princeton Tokamak Fusion Test Reactor

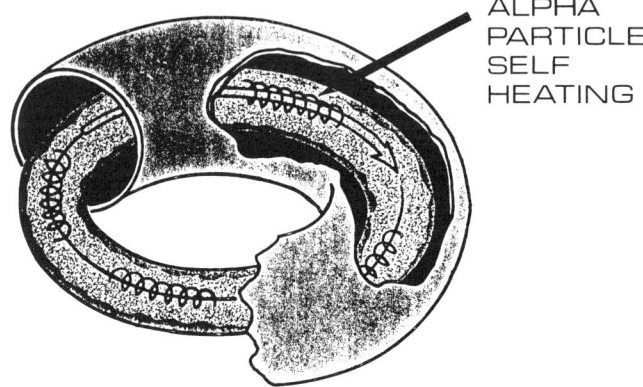

PLASMA HEATING AFTER IGNITION

Figure 18-7. Tokamak Fusion Reactor

the critical basic physics areas (see figures 18-8 through 18-14, and table 18-2).

General Atomics Doublet III Tokamak

The Doublet III device at General Atomic Company in San Diego is the world's largest operating tokamak and the first major fusion experiment to be conducted as a cooperative international endeavor under the terms of an agreement between the governments of the United States and Japan.

A major objective of the Doublet III experimental program is to achieve reactor-level plasma conditions in a hydrogen plasma (deuterium and tritium plasmas will not be used). This will serve to confirm the validity of the extrapolation of tokamak scaling laws to the reactor regime. The scaling law, empirically derived from experiments in a number of tokamaks, are predictions of plasma behavior in larger and more powerful machines. A tokamak fusion reactor will require a plasma that is approximately two to three times larger than the present plasma size in Doublet III. A modification that would substantially increase the size of the Doublet III plasma chamber is under consideration, but even the present size is adequate to confirm most physics aspects (except those involving alpha-particle confinement, heating, and scaling) of present reactor designs.

Table 18-1
Physics Requirements for a Tokamak Reactor

	Parameter	Description	Required Level
Basic plasma physics	$n_e \tau^e$ (plasma density × confinement time)	Energy confinement	(Lawson criteria) $2\text{-}3 \times 10^{14} \text{cm}^{-3}\text{s}$
Present experiments plus TFTR	Ion and electron temperature	Temperature	10 KeV
	β—beta	Plasma pressure as a percentage of magnetic-field pressure	4%
Minimum applied plasma physics requirements	Long-pulse power handling	Power-handling capability	$P > 100$ MW
	Long-pulse plasma equilibrium	Equilibrium maintainability	$T \geq 30$ s
Advanced tokamaks (FED/ETF/INTOR)	Long-pulse impurity control	Impurity control	$N_{\text{metal}} < 0.1\%$ $N_{\text{oxygen}} < 1\%$
Reactor plasma physics	High first-wall loading	Reactor efficiency	$2\text{-}3 \text{ MW/m}^2$
	Ash removal	Helium-ash control	$n_\alpha / n_e < 3\%$
EPR/DEMO/STARFIRE	Plasma fusion-power density	Power balance	2 MW/m^3

Table 18-2
The Doublet Program (GA)

Present D III Parameters	Upgrade Parameters
a = 45 CM	
b = 143 CM	
R = 145 CM	
B_t = 2.6 T	B_t = 4.0 T
I = 2 MA	I = 5 MA
	P_{AUX} = 18 MW NB
	Special Emphasis
	Configurational flexibility
	Intense neutral-beam heating
	High-beta optimization

An important objective of the original design of the Doublet III device was to produce magnetic configurations optimized in their ability to contain the plasma stably. A test of the expected stability to high plasma pressure awaits the experiments (just about to begin) with auxiliary heating, but two basic requirements have been demonstrated: the ability to control the detailed plasma shape and high plasma-current operation.

The ability of a magnetic configuration to isolate a plasma from the material walls is theoretically improved by elongating the cross-section of a tokamak plasma so that it is D-shaped rather than circular. The ability to produce such stretched cross-sections was an essential feature in the design of Doublet III. In Doublet III a special configuration can be set up in which the magnetic field expands the lower boundary of the plasma into the bottom half of the chamber (see figure 18-8); this configuration has been explored as a means of controlling impurities.

Plasma stability and energy confinement are improved as the plasma current is increased. The spatial distribution of the current in a typical circular plasma is generally peaked in the center—the current tends to be highest through a small circle at the center of the plasma cross-section. But the plasma can carry more current and resist sudden disruptions if the peaked plasma-current profile is flattened so that the outer plasma carries as much current as the inner plasma. One way of flattening the current profile is to elongate or enlarge the plasma cross-section, and experiments carried out in Doublet III have thus been able to yield record values of plasma current stably maintained. Currents of 1.5 MA in a "doublet" plasma and 1.0 MA in a "singlet" plasma have been achieved.

The energy confinement of these elongated plasmas has also been studied, and the results are consistent with the improvement expected theoretically. Confinement is improved about 40 percent in the ohmically

Status of U.S. Tokamak Effort

OPERATING DEVICE	PLASMA SIZE	RESULTS
PLT PRINCETON LARGE TORUS *Princeton Plasma Physics Laboratory*	$a = 40$ cm $R = 135$ cm $B_T = 35$ kG $I_p = 500$ kA	• NB HEATING $T_i = 7$ keV • $T_e = 4.5$ keV • ICRF HEATING $T_i = 2.5$ keV • ELECTRON TRANSPORT SCALING
ISX IMPURITY STUDIES EXPERIMENT *Oak Ridge Nat'l Lab.* 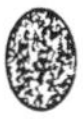	$a/b = 27/40$ cm $R = 93$ cm $B_T = 12$ kG $I_p = 120$ kA	• HIGH BETA $\beta \sim 4\%$ • IMPURITY REJECTION BY CO-BEAM
Alcator C *Massachusetts Institute of Technology*	$a = 16$ cm $R = 64$ cm $B_T = 90$ kG $I_p = 500$ kA	• HIGH DENSITY $n \sim 10^{15}$ cm^{-3} • HIGH $n\tau$ $3 \cdot 10^{13}$ cm^{-3} sec
Doublet III *General Atomic* 	$a = 45$ cm $R = 145$ cm $B_T = 26$ kG $I_p = 2$ MA	• EFFECT OF CROSS-SECTION SHAPE ON MHD STABILITY AND TRANSPORT • HIGH PLASMA CURRENT $I_p \sim 2$ MA
PDX POLOIDAL DIVERTOR EXPERIMENT *Princeton Plasma Physics Laboratory* 	$a = 40$ cm $R = 140$ cm $B_T = 24$ kG $I_p = 500$ kA	• DIVERTOR IMPURITY CONTROL • PERPENDICULAR NEUTRAL BEAM INJECTION $T_i \sim 2.5$ keV
TFTR TOKAMAK FUSION TEST REACTOR *Princeton Plasma Physics Laboratory* 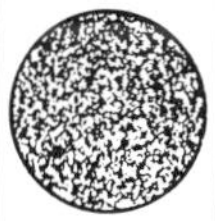	$a = 85$ cm $R = 248$ cm $B_T = 5.2$ kG $I_p = 2.5$ MA	UNDER CONSTRUCTION

Figure 18-8. Current U.S. Tokamak Experiments

heated plasmas but is expected to be better still with neutral-beam heating at high plasma density. A characteristic feature of tokamaks is the improvement of the energy confinement with plasma density according to a relation known as Alcator scaling (after a tokamak at the Massachusetts Institute of Technology). But at sufficiently high density the confinement begins to decrease with increasing plasma density as the ion losses become dominant over the electron losses. It is in this ion-dominated confinement regime that elongating the plasma cross-section can significantly improve the plasma confinement.

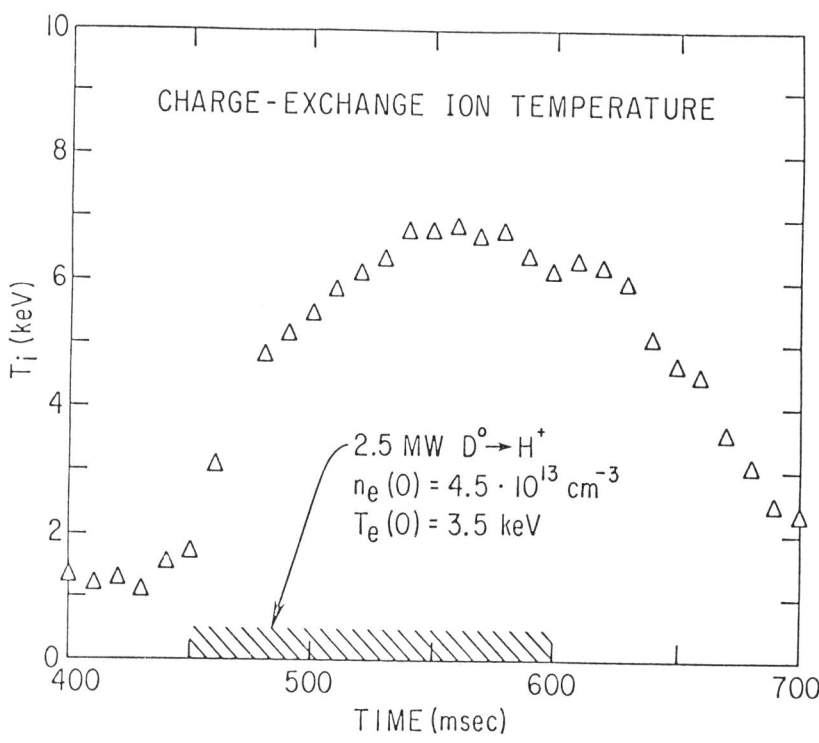

Figure 18-9. Neutral-Beam Heating in PLT

The improvement of the energy confinement with plasma density thus places considerable emphasis on operating at high plasma density. However, another tokamak empirical law (Murakami scaling) suggests that the maximum achievable plasma density in an ohmically heated discharge is directly related to the size of the toroidal magnetic field and inversely proportional to the major radius of the torus (distance from the center of the machine to the outer edge of the chamber); attempts to achieve higher density result in sudden plasma disruptions. Doublet III has achieved the largest value of plasma density relative to this scaling law of any ohmically heated tokamak. Devices with substantial auxiliary heating (PDX, ISX-b) have also been able to increase their operating density further relative to Murakami scaling.

Using special magnetic configurations, Doublet III has produced simplified magnetic-diverter configurations that pose more tractable engineering problems for reactor designers than the conventional diverter approaches for impurity control and ash removal (such as the internal

Status of U.S. Tokamak Effort

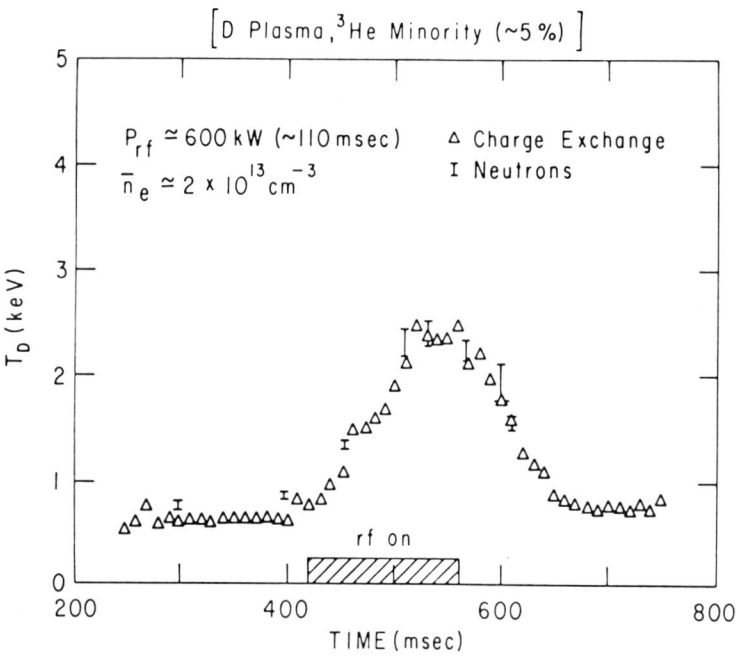

Figure 18-10. ICRF Heating on PLT

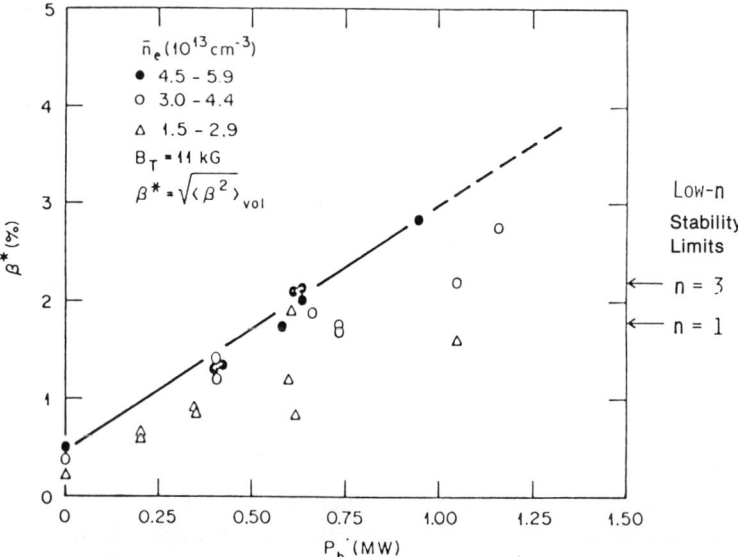

Figure 18-11. ISX-B Beta Values Exceed Expectations of Ideal MHD Theory

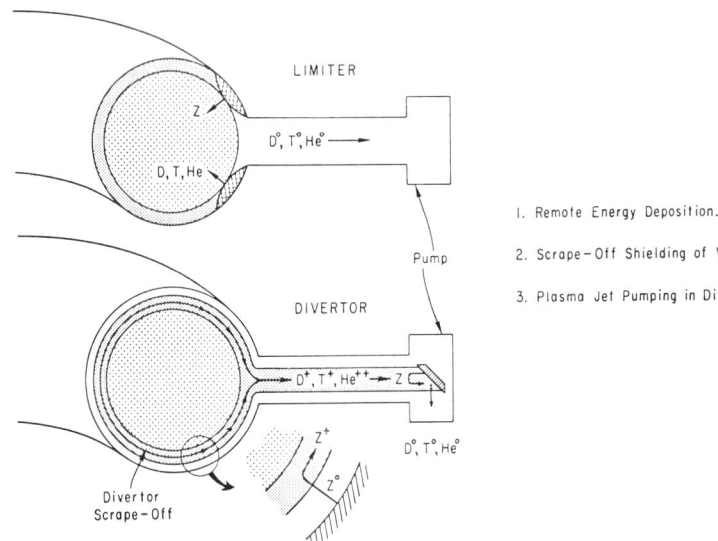

Figure 18-12. Comparison of Mechanical Plasma-Limiter Performance with That of PDX Magnetic-Divertor System

diverter coils of the Princeton PDX). Substantial reductions in impurity content in the plasma have been achieved using these simplified diverter configurations. Another favorable aspect of this diverter operation is the ability to force the edge region to radiate strongly, thus spreading the energy outflow from the plasma uniformly over the material container rather than concentrating it on diverter plates (see figure 18-15).

The Princeton TFTR

One important consequence of the very real progress in fusion research is that devices such as TFTR that look ahead to reactors tend to be very large and very expensive, and take a long time to build. For TFTR, for example, some six-and-a-half years will have passed from construction authorization to the earliest operation with hydrogen plasmas. Therefore, there is considerable potential that new developments from ongoing experiments could easily result in the need for drastic revisions before completion. In fact, the experimental results from smaller machines fully support the likely attainment of the basic TFTR objectives; and although this has led to a substantial increase in activity in the upgrade program called the TFTR Flexibility Modification (TFM) program, no significant changes in the base

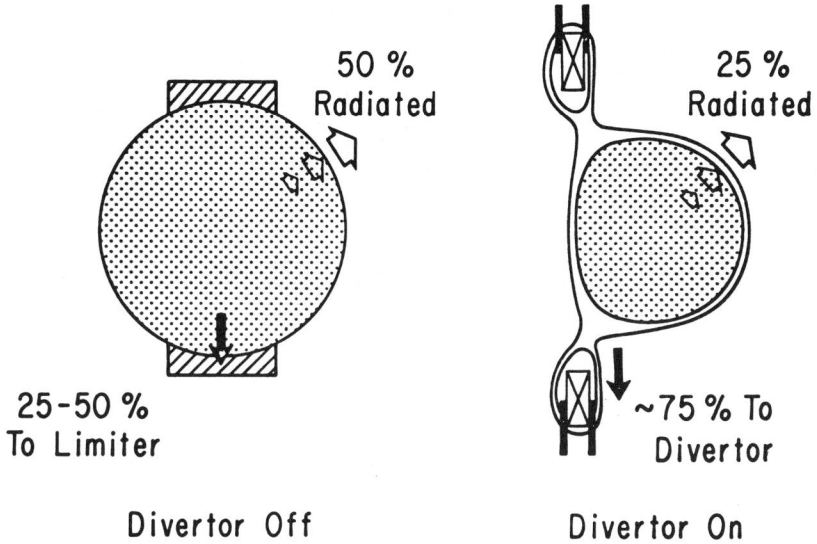

Figure 18-13. Performance of PDX Magnetic Divertor

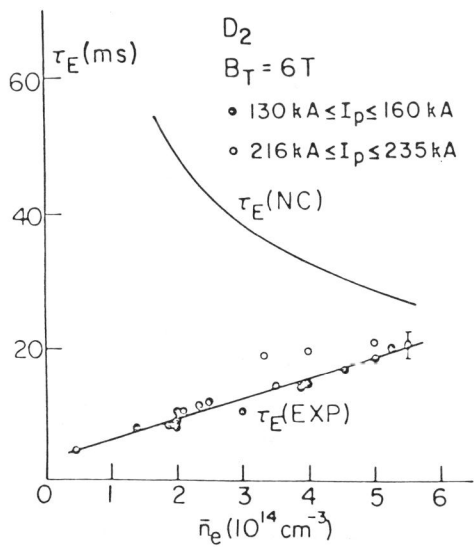

Note: For optimal plasma parameters, τ_E is close to classical expectation.

Figure 18-14. Energy Confinement Time on Alcator-A

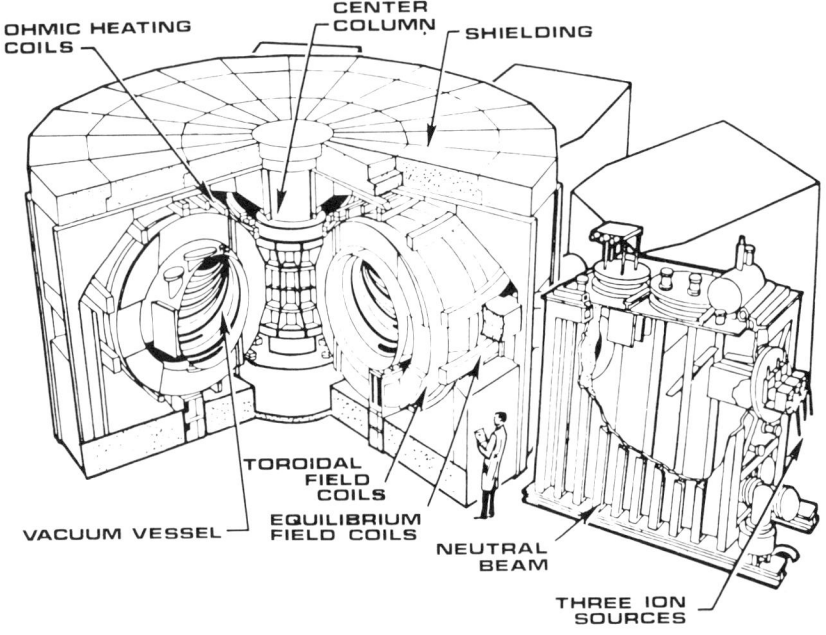

Figure 18-15. Artist's Conception of TFTR

TFTR are contemplated. As a result, the main physics and engineering objectives of the base design are mostly unchanged from those reported previously. These are summarized in figure 18-16. As indicated in the figure, the essential change provided by TFM is the increase in neutral-beam heating to move from a regime dominated by two-component fusion effects to one in which the fusion yield arises predominanty from bulk-plasma reactions, and where the reaction alphas contribute some 20 percent to the heating. These TFM effects come from the addition of a fourth beam line and the extension of the heating-pulse duration from 0.5 seconds to 1.5 seconds. An artist's conception of TFTR is shown in figure 18-20 to remind the reader of the general arrangement and some component details.

The major TFTR design parameters are shown in figure 18-17. The numbers for the base TFTR are essentially the original numbers as set in 1975; the indicated changes are those resulting from the TFM program. These changes are largely associated with the addition of power-supply modules to extend the pulse lengths (and in some cases the levels); the coil systems themselves were for the most part designed at the outset for the longer pulses. About half of the D-T pulses are expected to be at TFM

Study Physics of Reactor Grade Plasmas

$n\tau \sim 10^{13} - 10^{14} \text{cm}^{-3}$ sec.
$T \sim 5 - 10$ keV
DT POWER $\sim$ 1 Watt cm^{-3} — 18 MW Total
TFM $\left\{ \begin{array}{l} \text{Extend To Bulk Heating} \\ \text{Enhanced Alpha Effects} \end{array} \right\}$ — 40 MW

Advance Engineering Base

- High Power Neutral Beams 120 keV, 0.5 → 1.5 sec.
- Safe-Reliable Systems For Tritium Handling
- Remote Handling & Maintenance of Tokamaks
- Demonstrate Techniques For Handling High Pulsed Thermal Loads on Vessel Walls

Figure 18-16. TFTR Objectives

levels, leading to a radiation design level of 10^{22} neutrons per year (compared with 3.5×10^{21} for the base TFTR at 1,000 full power shots per year).

Figure 18-18 shows the layout of the TFTR facilities and how additional major experimental devices could be placed there at a later date, fully utilizing the auxiliary equipment and apparatus constructed for TFTR (a capital value of about $150 million in 1977 dollars, roughly half the total capital acquisition cost of TFTR). A prototype neutral-beam system has been developed by the Lawrence Berkeley Laboratory (in collaboration with the Lawrence Livermore Laboratory) for TFTR. TFTR can use six of these high-power beams, but only four have been authorized to date. TFTR is being built on a collaborative basis by the Department of Energy, laboratories, and industry. With Princeton as the systems manager, major elements of the program are being designed, procured, and/or installed by Ebasco Services, Inc. (the major industrial contractor); Grumman Aerospace; Giffels Associates; Lawrence Berkeley; and PPPL. The DOE has directly managed the construction of the conventional facilities. General Atomic is working with PPPL on first-wall and neutronic systems. A consortium similar to the one used on TFTR is being discussed for con-

Operation
    3×10^5 High-Power Pulses, 300 sec. Period
    4,000 Total, 1,000 per Year, 96 per Day of High
       Power DT Pulses – Half at TFM Levels
    Maximum of 10^{22} Neutrons per Year
    Maximum Tritium Charge ~ 400 Ci / Pulse

Tokamak
    Toroidal Field ...... 5.2 T   1.6 sec., 3.0 sec.(TFM)
    Ohmic Heating ....13.0 V.S.   1.0 sec., 3.0 sec.(TFM)
    Equilibrium Fields.. 0.42 T   1.0 sec
                          0.5 T   3.0 sec (TFM)
    Plasma Current .... 2.5 MA   3.0 MA (TFM)
    Vessel .... 1.1 M Minor Radius, 2.65 M Major Radius

Power System
    Main Supply .... 2 ac MGF, Delivers 4,500 MJ, 900 MVA
    AC → DC .... Transformer - Thyristor Convertors
    N.B. .... Transformer - Rectifier - Series Vacuum Tube

Figure 18-17. Major TFTR Design Parameters

struction of new major fusion devices. The significant management experience on this type of joint activity between government, laboratory, and industry on a major project like TFTR is as important for planning new major fusion systems as are the scientific and engineering results.

Other Large Tokamaks

In addition to the U.S. TFTR, three other major tokamak facilities are under construction: the European JET, the Japanese JT-60, and the Soviet T-15. These devices are all more ambitious than earlier ones in the countries involved and, together with TFTR, will complete the development of a formidable tokamak physics and engineering base from the mid-1980s to the early 1990s.

    TFTR is scheduled to come on the air in 1982, JET in 1983, JT-60 in 1984, and T-15 in 1985. Through annual and other special meetings there is a significant exchange of scientific and engineering information among all four project teams and a sincere desire by all international groups to preserve the complementarity of these four major facilities. In hydrogen plasmas all four can certainly do similar experiments to provide a cross-

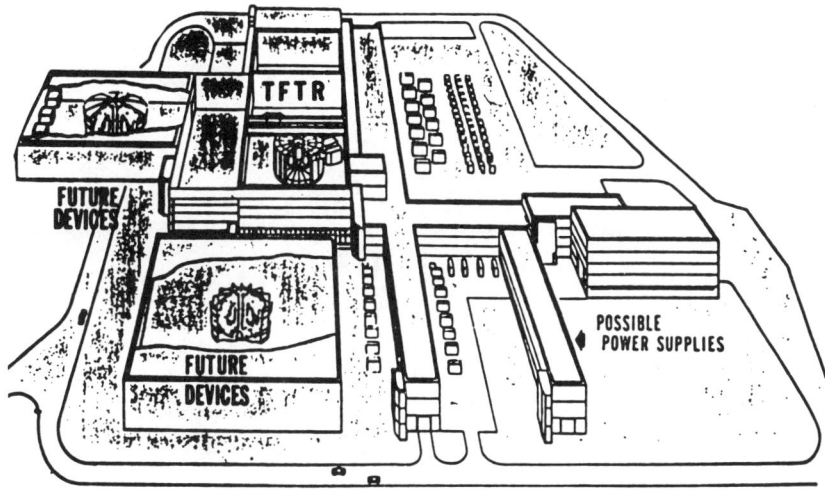

Figure 18-18. TFTR Site with Future Devices

check on each other's results; but only two of them will use thermonuclear fuel, only one has a magnetic diverter, and only one has a superconducting toroidal magnet system. A broad spectrum of information crucial to the development of criteria for a tokamak demonstration will emerge. It is crucial for future international endeavors that facilities not be exact duplicates of each other; the time it takes to build major facilities is so long that we need different major ones being constructed and operated simultaneously, each concentrating on different problems with a full sharing of the scientific results. The present collaborations among these four machines are indeed a step in the right direction. Figures 18-19, 18-20, and 18-21 compare the cross-sections and features of the four large tokamaks.

Summary of Progress on Tokamak Development

The U.S. tokamak effort has been thoroughly reviewed. Major reviews were held in March 1978 and March 1980. The first of these was chaired by Dr. John Foster of TRW and the second by Dr. Solomon Buchsbaum of Bell Telephone Laboratories. These review committees have had a significant effect on the shaping of plans and policies for future tokamak developments, which will be discussed later. The important aspects in tables

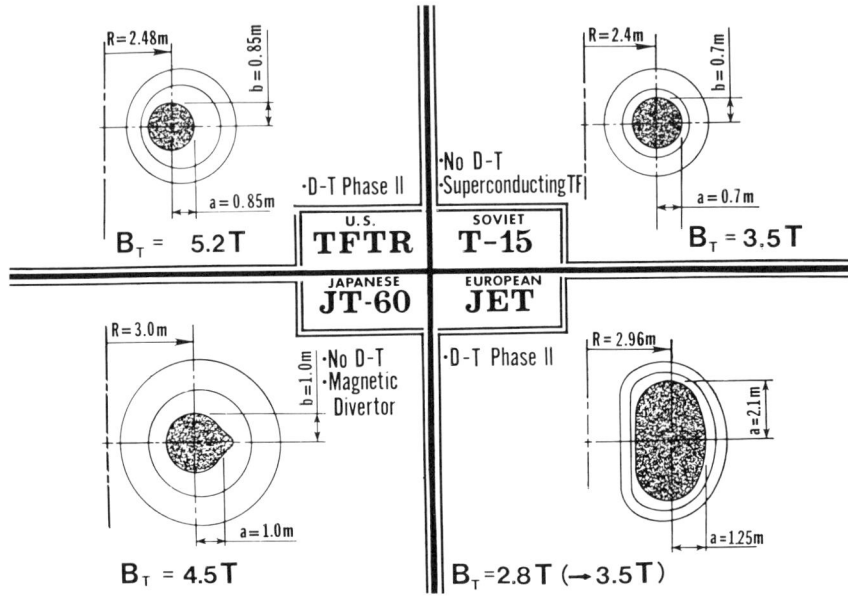

Figure 18-19. Comparative Cross-Sections of the Large Tokamaks

18-3 and 18-4, however, are to see, first, the progress in two years on three basic physics parameters, and, next, the development of new approaches. The news is all reasonably good, and we are beginning to see more choices available than ever before on the techniques that will improve the tokamak as a reactor candidate. Figure 18-22 shows the growth in the plasma size of Princeton devices over the last decade. The key points are that Q, the fusion-power multiplication factor, will have gone from an equivalent of about 10^{-4} to 10^0, or four orders of magnitude, in about a decade. During the same period plasma sizes will have increased by about a factor of ten, with no new major confinement problems showing up. The extrapolations required in Q and plasma size over the next ten years between TFTR and ETF are significantly less than those achieved over the last ten years.

Engineering Test Facility

For some two years, until recently, the tokamak fusion community (largely through the efforts of a joint laboratory-industry ETF design center located at ORNL) has been developing the conceptual specifications for a tokamak engineering-test facility that would be a major step forward in the area of engineering and particularly in nuclear engineering.

Status of U.S. Tokamak Effort

	United States TFTR/TFM	Europe JET	Japan JT-60	USSR T-15
Plasma Geometry	Circular $A = 2.9$	D-Shaped $A = 2.4$	Slightly Noncircular $A = 3.2$	Circular $A = 3.4$
Impurity Control	Getter Systems		Single-Nul Poloidal Divertor	
Plasma Temperature	10 - 20 keV	10 - 20 keV	10 - 20 keV	10 keV
$n\tau_E$	$\sim 5 \times 10^{13}$	$\sim 1 \times 10^{14}$	$\sim 5 \times 10^{13}$	$\sim 5 \times 10^{13}$
Pulse Length	1.0 - 1.5 sec	5 - 15 sec	5 - 10 sec	5 sec
Plasma Heating	$\sim$33 MW Neutral Beams	10 MW Neutral Beams, 15 MW Radio Frequency	15 MW Neutral Beams, 10 MW Radio Frequency	9 MW Neutral Beams, 6 MW ECRH
Energy Multiplication	$Q_p \sim 2$	$Q_p \sim \gg 1 - 5$	None	None
Plasma Current	2.5 - 3.0 MA	5.0 MA	2.8 MA	2.0 MA

Figure 18-20. Four Large Tokamaks Planned for the 1980s: Physics Aspects

	United States **TFTR/TFM**	Europe **JET**	Japan **JT-60**	USSR **T-15**
Toroidal-Field Coils	Circular Resistive	D-Shaped Resistive	Circular Resistive	Circular Superconducting
Power Source	MG-Flywheel	MG-Flywheel and Grid	MG-Flywheel and Grid	Grid
Ohmic Transformer	Air Core	Iron Core	Air Core	Iron Core
Poloidal-Field Coils Position	External to TF Coils	External	Internal	External and Liquid Nitrogen Cooled
Tritium Usage	In Regular Operation	Late in Machine Life	None	None
Shielding	Igloo Around Machine	Building Only	None	None
Remote Handling	Some Maintenance Operations	Some Maintenance Operations	None	None
Blanket Tests	Test Stations for 1 × 1 × 1 m Modules	None	None	None

Figure 18-21. Four Large Tokamaks Planned for the 1980s: Technology Aspects

Table 18-3
Summary of Recent Progress in Heating, Stability, and Confinement

	Foster Committee Era (March 1978)	Buchsbaum Committee Era (March 1980)
Neutral-beam heating	$\Delta T_i \sim 1$ keV. $\Delta T_e \sim$?	$\Delta T_i \sim 5.5$ keV ($\rightarrow 7.0$) recent PLT $\Delta T_e \sim 1.5$ keV ($\rightarrow 4.5$) recent PLT
Beta stability limits	B★ < 1%, no contact with theory	B★ ~ 3% (4% recent ISX) Better than theory
Energy confinement	$n\tau_E \sim 10^{13}$ CM $^{-3}$SEC, Doubts about scaling to collisionless range	$n\tau_E \sim 3.10^{13}$ CM $^{-3}$SEC, (ALCATOR) Confinement improves in collisionless range

Note: Tokamak experiments have now demonstrated that a plasma of ETF/FED size can surmount the three traditional barriers to the realization of controlled fusion power.

Figures 18-23 and 18-24 show one of the significantly improved tokamak design features that was incorporated into the ETF. The need for remote maintenance on the entire tokamak coil, structure, and vacuum systems (as is the case for TFTR) has long been a drawback when considering ultimate tokamak reactor applications. As these figures show, when the tokamak is large enough, with a particular dimensional aspect ratio one can consider building a tokamak reactor in which the coils, their structural support, many elements of the vacuum chamber, and so forth are fastened together for the life of the machine and shielded from the neutrons with

Table 18-4
Summary of Recent Progress in Tokamak Improvements

	Foster Committee Era (March 1978)	Buchsbaum Committee Era (March 1980)
Convenient heating	No significant success with RF	ICRF $\Delta T_i \sim 1.7$ keV (recent PLT) ECRH $\Delta T_e \sim 0.5$ keV (ISX)
Impurity control	Limiter satisfactory, no divertor experience	Improved limiters, Initial divertor results
Quasi-steady-state operation	No pellet fueling, no RF-current drive	Pellet fueling success (ISX) "Physics" demonstration of RF current drive (PPPL)
Simplified mechanical structure	Low-access designs	Improved β prospects, high-access designs

Note: Ongoing tokamak experiments and engineering studies have begun to develop the improvements needed to make the tokamak (EPR/DEMO) a commercially attractive reactor candidate.

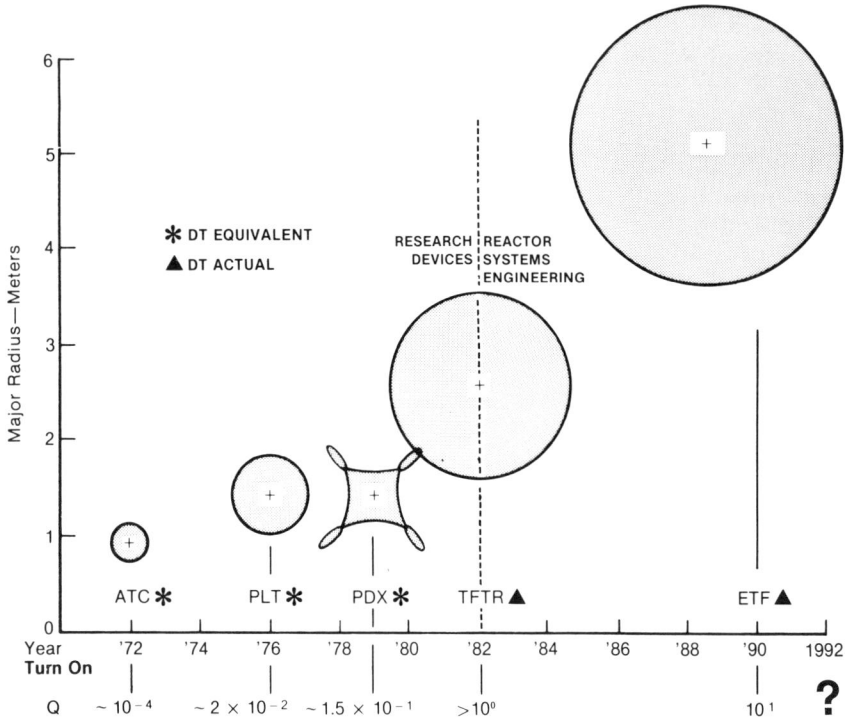

Figure 18-22. Tokamak Development

fixed in-place shielding. The blankets, first-wall components, and so on are then moved in and out of the reactor chamber through drawers that slide between the coils. Only these drawer modules need to be remotely maintained, a significant advantage over linear systems or smaller toroidal systems. The ETF and Starfire (to be discussed later) have this feature. Figures 18-25 and 18-26 compare ETF with TFTR. In these comparisons, one should note that the physics extrapolations are generally much less than the technology ones, except in the case of neutron production and stored energy.

It has recently been decided that the ETF is too ambitious for the time being, and planning for its construction has been suspended.

Center for Fusion Engineering

As a result of the work of the Buchsbaum ERAB committee, a special Fusion Panel appointed to advise the U.S. House of Representatives Science

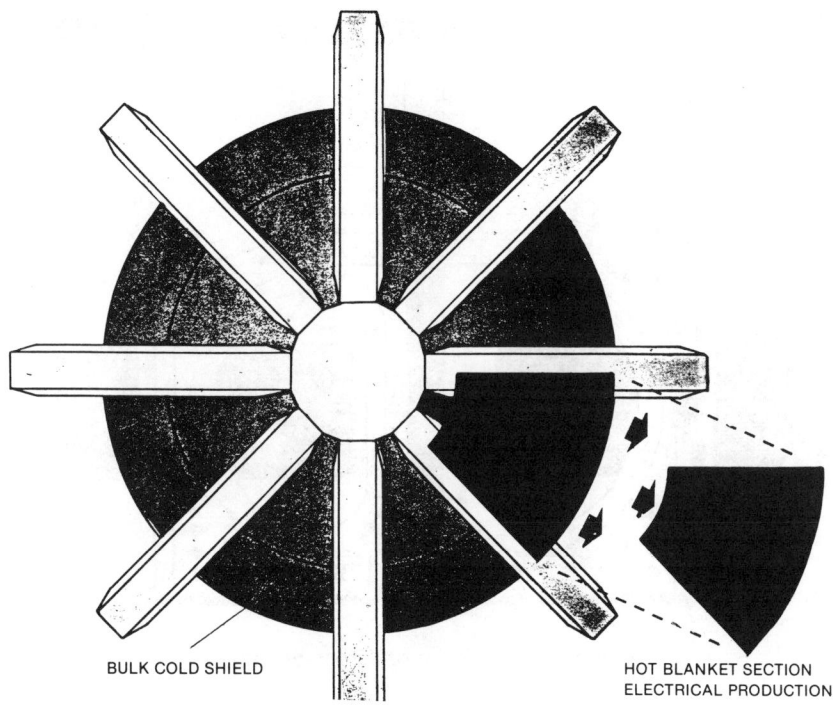

Figure 18-23. Typical Fusion-Reactor (Plan View)

and Technology Committee, the DOE, the institutions engaged in fusion energy development, and several industrial groups (including the Atomic Industrial Forum and Fusion Power Associates), a consensus was developed that clearly indicated that fusion-power development was ready to enter into the engineering phase. This effort culminated in the passage by the U.S. Congress of the Fusion Energy Act of 1980, after similar bills (introduced by Representative McCormack in the House and Senator Tsongas in the Senate) had passed both houses. The act, signed into law by President Carter, calls for, among other things, the establishment of a U.S. Center for Fusion Engineering, the construction of a Fusion Engineering Device, the provision of a demonstration fusion reactor by the year 2000, a doubling of the funding applied to fusion-energy development over the next several years (at a total cost of about $20 billion by 2000) in order that the viability of fusion power over the next twenty years can be established. The DOE is to report to Congress by July 1981 on its plans for the establishment of the Center for Fusion Engineering (CFE).

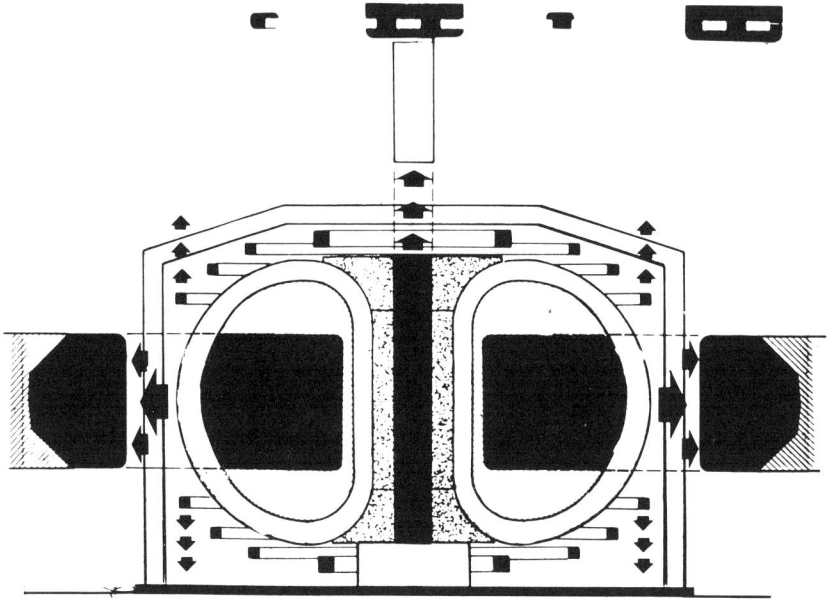

Figure 18-24. Reactor Cross-Section

Table 18-5 gives the author's impression of the CFE mission. Although there are no official DOE statements yet about the CFE, it is clear that if it is set up in accordance with the Fusion Energy Act of 1980, the CFE will play a major role in the shaping of fusion-energy development over the next generation.

Table 18-6 shows the very preliminary features of the Fusion Energy Device, which is now being looked at by the former Engineering Test Facility Design group at ORNL. The FED is intended to be a major step in engineering beyond TFTR, but not as significant as the ETF, particularly in the areas of high neutron flux and overall availability. The FED design criteria and overall requirements are being established by a National Technical Management Board.

Tables 18-7 through 18-11 are self-explanatory and give the author's views on the experimental phases of FED, what it can be expected to accomplish, and what will not yet be accomplished at CFE even after FED is completed. A key feature of the planning of CFE and FED must be the successful integration of the work of the present fusion-energy-development institutions. The experimental programs, the R&D, and the talents of the present institutions must be fully utilized over the lifetime of the CFE, particularly in the design, construction, and testing of FED in a full partnership role if the venture is to be a success. Fusion-power development is not at the stage when it can be completely turned over to industry. The established

Status of U.S. Tokamak Effort

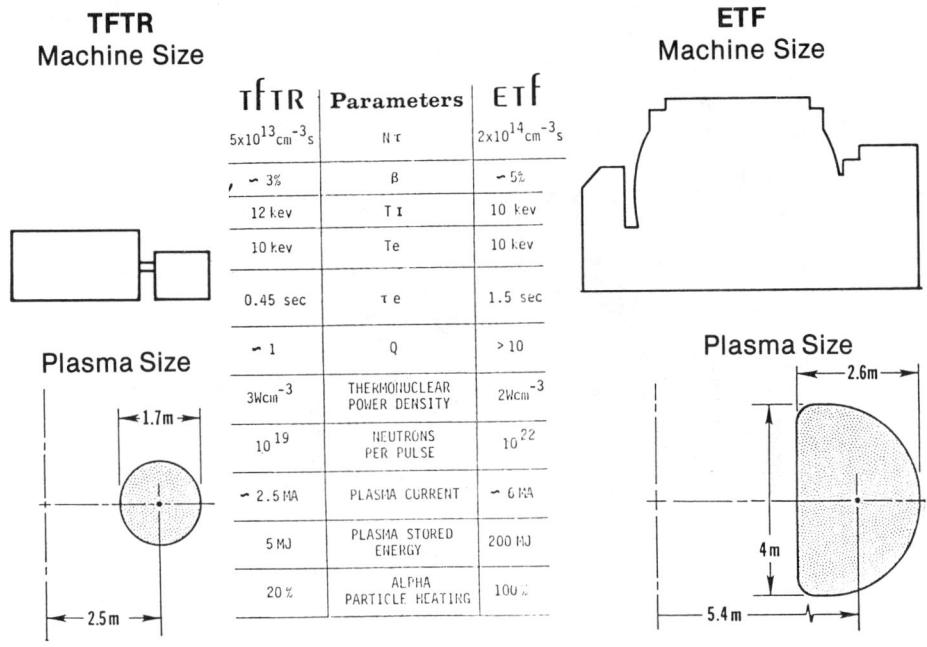

Figure 18-25. Physics Comparisons between TFTR and ETF

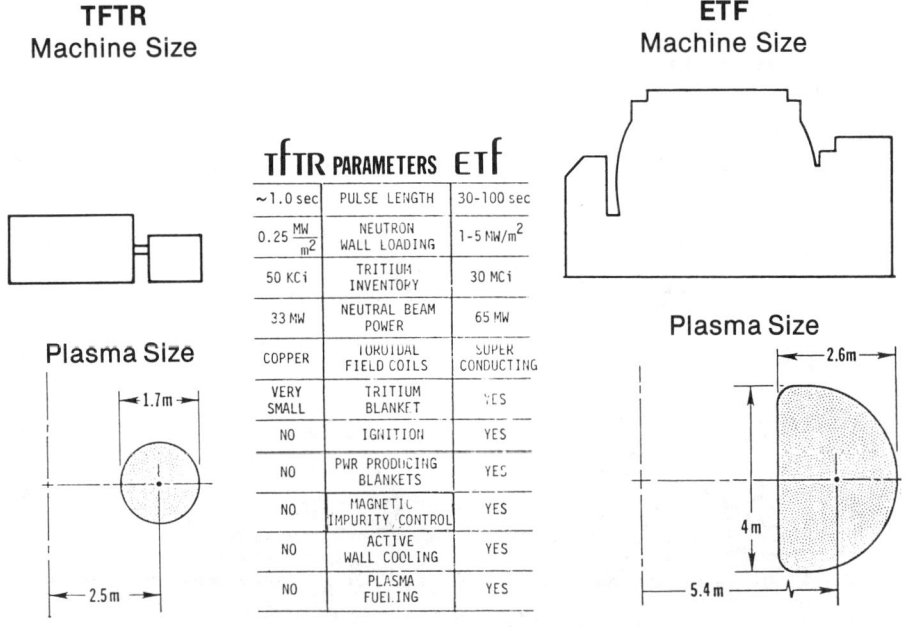

Figure 18-26. Technology Comparisons between TFTR and ETF

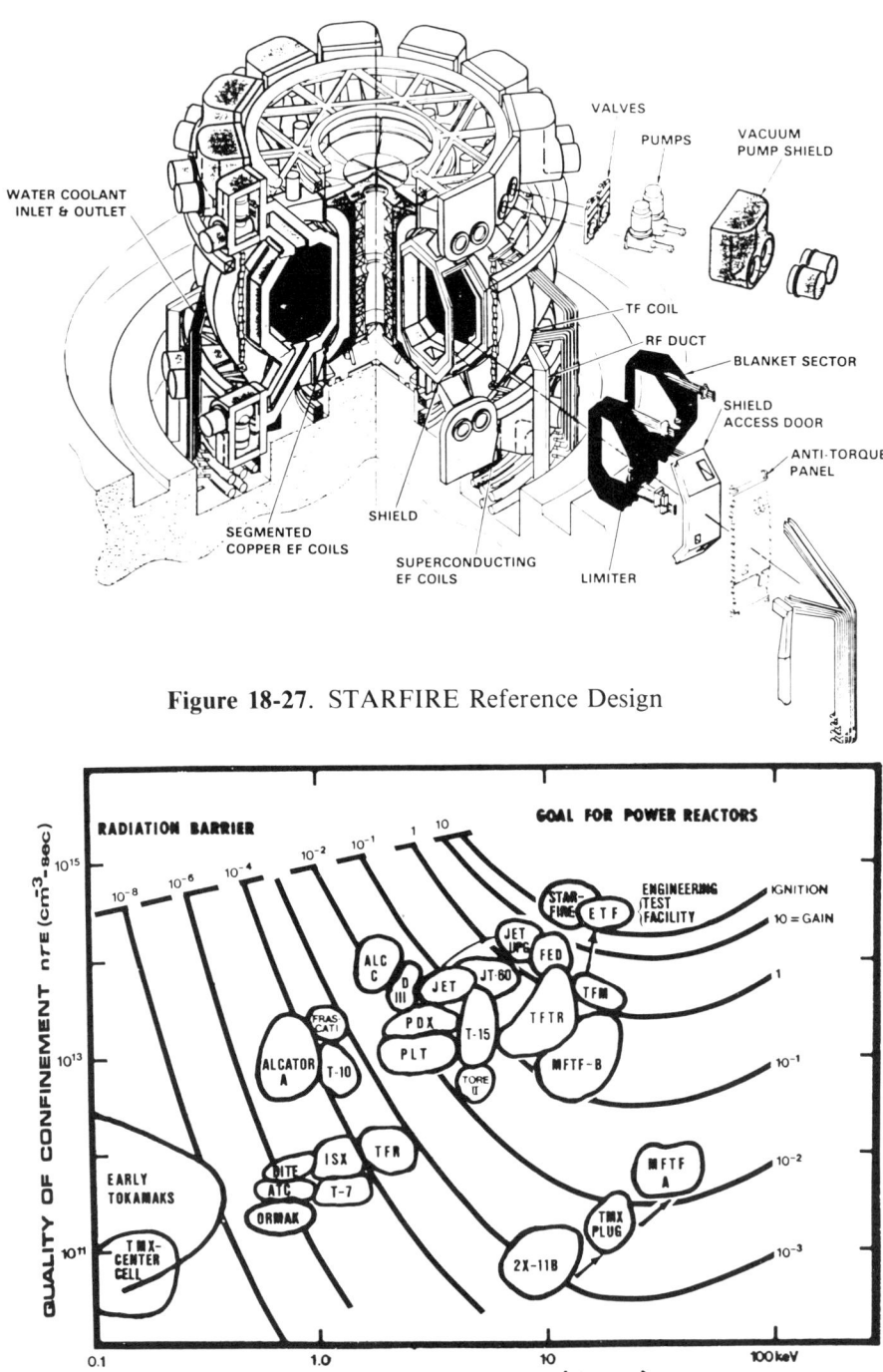

Figure 18-27. STARFIRE Reference Design

Figure 18-28. Progress toward Fusion Power

Table 18-5
A Proposed CFE Mission

Foster and demonstrate engineering development of fusion power from present through a demo in 2000.
Establish a strong multidisciplinary core group to accomplish the CFE mission from inception of FED through operation of the demo.
Participate with DOE and other fusion institutions in the development of a national (or even international) fusion-power development program.
Assist DOE and other fusion institutions in assignment of prime FED or CFE subsystem responsibilities to existing fusion R&D institutions.
Establish with DOE and non-tokamak fusion institutions FED and CFE objectives consistent with and supportive of other than tokamak fusion-device development.
Through an architect/engineer/contractor build and operate the CFE facilities that will house the FED and other devices necessary to accomplish CFE mission.
Establish performance criteria for FED; its first wall, blanket, and shield modules; and other critical R&D items; and select prime industrial subcontractors to design and construct them.
Establish comprehensive programs for fusion-reactor safety and environmental-control systems.
Develop a comprehensive fusion-reactor control-simulation system and use it on FED.
Develop and demonstrate fusion-reactor diagnostic systems.

capital and human resources must be marshaled with the broader and essentially different industrial resources, with equal emphasis on both, for the effective prosecution of the next engineering phase of fusion energy development. The TFTR is an early example of this type of partnership, and there is much to be learned from that experience.

Table 18-6
Probable Fusion Engineering Device Features (Less Ambitious Than ETF)

Burning D-T tokamak
Ohmic stabilizing current
Beam- or RF-driven (Ignition perhaps an upgrade)
$n\tau\ 7 \times 10^{13}$
$\beta \sim 4\%$
$T_i = T_e \sim 10$ keV
$I \sim 7$ Meg Amps
8 Tesla toroidal field
200–300 MW thermonuclear power
$Q \sim 5$
0.5 MW/M^2 wall loading
Neutron flux 5×10^{17} N/cm^2/sec
Pumped limiter (probably no magnetic divertor)
25–100 = sec Pulse
5×10^4 D-T shots

Note: Major trade studies underway.

Table 18-7
Typical Phase I Fusion Engineering Device (FED) Experimental Program: Demonstration of Pulse to Pulse Reproducibility of Plasma System

Plasma profile control over FED long pulse
  Stability
  Current drive
  Current density
  Temperature
  Position
  Profile Uniformity
Active feedback system
Disruption control
Programmed start-up and turnoff
Safe handling of plasma stored energy in a controlled plasma shutdown

Note: The responsibility for developing and designing these systems for FED could be assigned to one or several existing tokamak laboratories, which would demonstrate the feasibility of the system proposed in existing or new facilities at their site in accordance with design criteria, interface control, cost and schedule, and so forth established by CFE for FED.

Table 18-8
Typical Phase II FED Experimental Program Demonstration of First Wall and Plasma Composition Systems

Plasma shaping and inner vacuum vessel protection
  Uniform thermal wall loading
  Limiter design
  Protective plate and/or armor design
  Particle exhaust and/or vacuum-pumping channel design
  Thermal deposition and fatigue considerations
Plasma heating
Plasma fueling
Plasma impurity control
Helium-ash-removal system

Note: The responsibility for developing and designing these systems for FED could be assigned to one or several existing tokamak laboratories, which would demonstrate the feasibility of the system proposed in existing or new facilities at their site in accordance with design criteria, interface control, cost and schedule, and so forth established by CFE for FED.

Table 18-10
Major Generic Issues to Be Resolved by Fusion Engineering Device (FED)

Select promising first wall and shielding material and configurations including resolution of thermal and fatigue effects.
Select promising tritium-breeding blanket module configurations.
Select promising power-producing module configurations.
Maintainability of modular fusion-power reactor systems.
Environmental and safety requirements for a fusion-reactor site.
Plasma-burn requirements (fueling and exhaust) of a fusion-reactor system.
Operation of a prototypical fusion-reactor system.
Reliability of superconducting magnet systems for fusion-reactor applications.
Reliability of fusion-reactor plasma heating, control, and diagnostic systems.
Handling and disposal of radioactive materials in a fusion-reactor environment.
Handling and control of a fusion-reactor-level tritium inventory.

Table 18-11
Major Issues to Be Resolved at CFE after FED

Tokamak or non-tokamak EPR/demo
If tokamak EPR/demo:
   Steady state or pulsed
   Current-drive system
   Plasma-heating system
   Size (TF field and plasma current)
   Impurity-control system
   Ignition or driven
   Availability
   Duty factor
Reactor-power output on basis of utility needs
Manufacturability of different fusion-reactor devices
Economics of different fusion-reactor devices
Availability requirements
Reliability requirements
Quality-assurance requirements
Tritium requirements
Safety and environmental requirements

Table 18-12
Key Features of STARFIRE

Steady-state plasma operation

Lower hybrid RF for plasma heating and current drive

ECRH-assisted start-up

Limiter/vacuum system for plasma-impurity control and exhaust

All superconducting EF coils outside TF coils

Vacuum boundary at the shield, mechanical seals

Total remote maintenance with modular design

Water-cooled, solid tritium breeder blanket with stainless steel structure

All materials outside the blanket are recyclable within 30 years

Less than 0.5 kg of vulnerable tritium inventory

Minimum radiation exposure to personnel

Conventional water/steam-power cycle with no intermediate coolant loop and no thermal-energy storage

---

Starfire

Table 18-12 and figure 18-38 show the features and artist's conception of the Starfire tokamak fusion-reactor design. This design was accomplished by Argonne, with the industrial participation of McDonnell-Douglas and Parsons Engineering. For design purposes, it is assumed that Starfire is the tenth production reactor and that it therefore can be designed with more optimistic promises (such as steady-state operation rather than pulsed operation), allowing for the fact that present-day issues can be resolved by that time. It has also been designed to have adequate availability and maintainability features as a result of significant input from a utility advisory committee.

Observations

Table 18-13 gives my observations on the state of tokamak development, and figure 18-28 indicates the progress being made toward the development of a tokamak reactor and shows how far we still have to go. My observations, though consistent with those of the Buchsbaum panel and the Fusion Engineering Act of 1980, reflect a concern about the danger of the program's slipping back into R&D. The commitment to engineering must come soon enough and be broad enough, with adequate early funding, if the goals of the Fusion Energy Act are to be achieved. Overemphasis on the least expensive next step (assuming that it must be a D-T device) could

**Table 18-13
Observations**

Scientific progress on tokamaks justifies major step beyond TFTR.

Technology base intrinsic to present tokamak facilities not adequately exploited; operation of present devices should be extended for solution of plasma engineering and first-wall problems.

Near-term emphasis on improved tokamak-plasma-physics performance of FED could delay initiation of sound fusion-engineering phase with CFE and FED.

CFE with broad responsibilities for program integration should be established as soon as possible.

Mission orientation of program should be reinforced and strengthened by assignment of some key technical responsibilities for FED to present institutions.

Improvement programs on TFTR, Doublet III, and other existing tokamaks, and construction of new advanced tokamaks, may be required to demonstrate FED credibility and to provide design criteria for a tokamak EPR or demo.

A front-end commitment not to compromise engineering, maintainability, and availability features of FED in favor of higher-performance plasma-physics parameters should be made.

---

seriously jeopardize the achievement of the demonstration by the year 2000. The design of the FED should be done only after the requirements for the demo are sufficiently established to highlight the objectives of intermediate devices. Similarly, an assessment should be undertaken of the suitability of present devices and test stands to resolve some of the crucial demo requirements. If the next step is either too big (requiring years of R&D before construction) or too small (insufficient performance to resolve key engineering issues), then the demo might be seriously delayed, since it takes almost ten years before a major device can be turned on.

19 Nuclear Fission as a Global Energy Resource

Robert J. Creagan

History

Oklo Reactor

Nuclear fission first produced large amounts of energy on this globe, according to available information, about 2 billion years ago at Oklo, now in the Republic of Gabon in Africa. About 10^{11} kwh (0.3 Q) of energy were produced during a million-year period, which involved 500 tons of uranium. As one indication of how long ago that was, figure 19-1 shows that Africa was adjacent to South America at that time. Another time indicator is that it was 2.8 half-lives (0.7 10^9 years) ago for uranium-235, so natural uranium enrichment was 3.5 percent at that time. The earlier time made Oklo's natural uranium fuel similar to today's water-reactor fuel, that is, 2.9- and 3.8-percent enrichment for 33,000 and 45,000 MWD/MTM burnups, respectively.

Today's 3,400-Mt (1,125-MWe) pressurized-water reactor produces 10^{11} kwh of heat in 5 years from 250 tons of 3 percent uranium, which is about twice the energy per ton at Oklo. Temperature and pressure at Oklo were similar to today's PWRs. The fact that the temperature was above 600°F, which is required for radiation-track annealing, can be inferred from lack of radiation-damage tracks, in spite of a neutron fluence of more than 10^{21} neutrons/cm². To retain water required for neutron moderation, saturation pressure of 2,000 psia is required at 636°F. Such a pressure was available from hydrostatic sources because Oklo was at least 5,000 feet underground at that time. Groundwater presumably concentrated the uranium and periodically was percolated out of the reactor over a period of a million years.

The Oklo reactor is a fascinating scientific detective story, and the first evidence was a tenth of a percent difference in the uranium -235/238 ratio. Table 19-1 shows the ratio as low as 0.44 percent in the reactor volume. Uranium concentration was 40-60 percent at Oklo, compared with about 80 percent for today's reactors. Fission products from Oklo, as shown in figure 19-2, are still concentrated in a limited volume after 2 billion years. The site was discovered in 1972, and since then the fission products and actinide isotopes in the vicinity have been studied exhaustively. This infor-

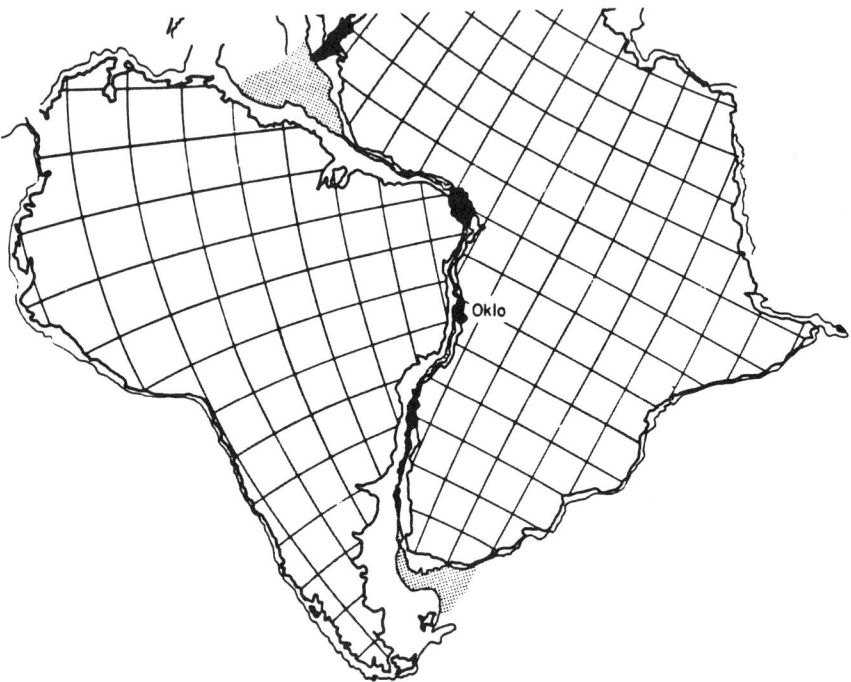

Source: From *Yearbook of Science and Technology*, P.K. Kuroda, 1979. Used with permission of McGraw-Hill Book Co. Copyright © 1979, McGraw-Hill.

Figure 19-1. Oklo Reactor 2×10^9 Years Ago

mation, although site specific, is available to nuclear-waste-disposal investigators—Africa moved, but the reactor ashes stayed together.

Because water concentrates uranium and can serve as a neutron moderator, other natural uranium reactors probably have operated. For example, Canada now evaluates its rich uranium-ore deposits for evidence

**Table 19-1
Reactor Data**

	Oklo	*1125 MWe PWR*
Uranium (tonnes)	>500	90
(W/O)	40-60	80
Burnup (MWD/MTM)	<8,000	33,000
Enrichment (%)	~3.5	2.9
Temperature (°C)	>300	300
Pressure (psia)	>2,000	2,250

Nuclear Fission

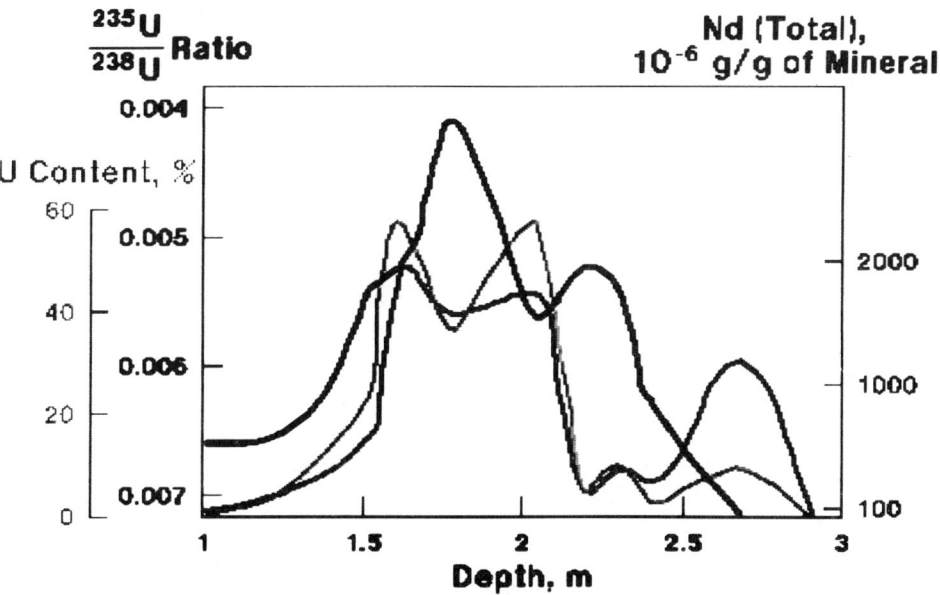

Source: From *Activités Scientifiques et Techniques*, Commissariat a l'Énergie Atomique, Paris, 1976. Used with permission.

Figure 19-2. Uranium Contents, U-235/U-238 Ratios, and Neodymium Isotopes in the Oklo Reactor Site

of fission activity. The time window conducive to such reactor operation is thought to be bounded by the first availability of photosynthesis, which might have provided chemicals for uranium concentration about 2 billion years ago, and by the time natural uranium was no longer enriched enough to go critical in a dirty-water moderator. For example, 1-percent enrichment was available 5 million years ago, Oklo happened because nature supplied all the ingredients, and coincidence brought them together at the right time.

Man-Made Reactors

In 1942 nuclear-fission reactors were reinvented by humans searching for a device to release nuclear energy. Early applications were motivated by military objectives, which included weapons and naval propulsion. Today's motivations are economic, and the only major application is as an energy source for large scale (>300 MWe) generation of electric power. In the last ten years only one power reactor (Fort St. Vrain, 342 MWe) less than 500

MWe has been built in the United States. There are other relatively minor applications, such as heat for building or process heat, but they do not use much uranium at present.

Energy Resources

Impact

Energy today truly has global impact. More specifically, the most desirable energy resource, oil, has become the most powerful economic force on earth. The presence or absence of petroleum resources in a country dictates economic viability, balance of payments, and global political and military stature. Hence the world seeks to substitute other energy resources to reduce oil consumption.

World nonrenewable energy resources and recoverable amounts are shown in table 19-2. A major message inherent in the table is that both the type of energy resources and the quantity of energy available to man, as a substitute for oil, are determined by recovery and utilization technology, in addition to discovery of the natural energy resource.

Fossil Resources

Conventional liquid oil fuels, so important for transportation, can be increased by the use of heavy oil, tar sands, and shale oil to 7,536 Q as technology and environmental laws permit. Conversion of coal to a liquid hydrocarbon, based on available technology at 66-percent energy efficiency, would make an additional 40,000 Q available as an oil substitute.

Fission-Water Reactors

Nuclear fission as a global energy resource, when limited to water-reactor technology, is comparable to global petroleum resources. An 1,125-MWe light-water reactor (LWR) can produce 3,411 MWt and 2 Q during thirty years' operation. This assumes no reprocessing, an average capacity factor including refueling of 65 percent, and use of 4,632 tons of natural uranium feed to provide enriched uranium fuel for a burnup of 33,000 MWD/MTM.[1] This is equivalent to fissioning 0.58 percent of the uranium mined.

Table 19-2
World Nonrenewable Energy Resources and Recoverable

	Resource	Q	Percentage	Recoverable—Percentage		Q	Percentage
Conventional petroleum (10^9B)	1,661	9,966	3.3	646^a	39%	3,876	5.5
Heavy oil (10^9B)	2,270	13,620	4.6	227^b	10%	1,362	1.9
Tar sands (10^9B)	964	5,784	1.9	193^c	20%	1,158	1.6
Shale oil (10^9B)	3,260	19,560	6.5	190^d	6%	1,140	1.6
Natural gas (10^{12}ft^3)	8,500	8,500	2.8	$2,519^a$	30%	2,519	3.6
Coal (10^9T)	12,682	241,708	80.9	$3,171^e$	25%	60,430	85.8
Total Q		299,138	100 %			70,485	100 %
Uranium—5 10^6T—used in light-water reactors	2,000 Q						
Uranium—used in breeder reactors				200,000 Q			
Lithium—9 10^6T—used in D-T fusion reaction				200,000 Q			

[a] Published reserves.
[b] Assumes 2 10^{12} barrels in Orinoco Oil Belt with 10% recovery.
[c] Based on surface mining (50%) plus in site processing (50%) of known deposits.
[d] Based on using present technology and considering only the higher grade, more accessible deposits.
[e] 25% recovery of resource is assumed, although reserves are published as 786 10^9T (6% of resources). Recoverability has been estimated as high as 50%.

Assuming 5×10^6 T of uranium available means a 2,150-Q energy resource. Without reprocessing, a 10-percent increase in available energy is possible by improved LWR fuel cycles.

An additional 10-percent increase in useful energy from natural uranium is available by use of heavy-water reactors. Candu reactors, at a natural uranium-fuel burnup of 7,500 megawatt-days per metric tonne of metal—(MWD/MTM) and 30.5-percent thermal-to-electric conversion efficiency—produce 2.3 megawatt-days electric per kilogram of natural uranium (MWDe/KgU$_N$) compared with 1.9 or 2.1 (MWDe/KgU$_N$) for the LWRs described earlier.

If reprocessing, which is now forbidden in the United States, is allowed, plutonium and uranium-233 recycle are possible. A LWR that produces 0.6 plutonium atoms per uranium-235 atom destroyed, might have 0.2 plutonium atoms available for recycle when fuel is discharged. Depending on the reactor in which it is recycled, 10-20 percent more energy might be obtained from the plutonium.

The thorium-uranium-233 cycle is feasible and has been used in both light-water and heavy-water reactors, as well as in the high-temperature gas-cooled reactor. Because uranium-233 does not exist in nature, it must be produced by neutrons from the uranium-235 fission process. The increase in total energy from uranium mined, by use of the thorium-uranium-233 cycle, might be doubled but would depend on the reactor in which the thorium is produced and recycled.

The 5-million-ton assumption for free-world uranium is based on table 19-3 (4,367.9 $\times$ 10^3 tonnes) rounded off because of data uncertainty. Table 19-3 data were referenced by the National Academy of Science in their 1979 book, Energy in Transition, 1985-2010.

The highest estimate for free-world uranium (6.6-14.8 million tonnes) was quoted by the NEA/IAEA Steering Committee Group on Uranium Resources published in World Uranium Potential: An International Evaluation by OECD in 1978.[2] The group's comment on the estimate was that even if such speculative resources were to exist, there would be no guarantee of their discovery or that they could then be made available. Uranium as a resource was thus taken at 5×10^6 tons.

Fusion-Fission-Coal

Coincidentally, 200,000 Q is the energy contained in total coal resources and available from uranium used in breeder reactors, which will be discussed later. It is also the energy potentially available from deuterium-tritium (D-T) fusion reactions, assuming 9 million tons of lithium (67,500 tons

Table 19-3
Estimated World Resources of Uranium Recoverable at Costs up to $130 per Kilogram as of January 1977 (10^3 tonnes)

Region	Reasonably Assured Resources	Estimated Additional Resources	Total
North America	825.0	1,709.0	2,534.0
Western Europe	389.3	95.4	484.7
Australia, New Zealand, and Japan	303.7	49.0	352.7
Latin America	64.8	66.2	131.0
Middle East and North Africa	32.1	69.6	101.7
Africa south of Sahara	544.0	162.9	706.9
East Asia	3.0	0.4	3.4
South Asia	29.8	23.7	53.5
World (except communist countries)	2,191.7	2,176.2	4,367.9

Source: *World Energy Resources, 1985-2020.* Copyright World Energy Conference, 1978. Published for WEC by IPC Science & Technology Press, Guildford, U.K., 1978. Pg. 116. Used with permission.

of lithium—6 at 7.42 isotopic percent) available, and 100,000,000° temperature confinement technology. For unlimited fusion energy, D-D fusion-reaction technology, with 300,000,000° ignition-temperature technology, is required.

Nuclear Applications

Oil Replacement

Employment of nuclear fission as a replacement for oil depends on temperature, minimum economic size, mobility of the oil application, and the required fission reactor. Maximum temperatures available from reactor working fluids are approximately 600°F for water reactors, 1,000°F for sodium reactors, and 1,500°F for gas-cooled reactors (1,000°C is a target). Minimum economic size depends on the application—for example, 300 MWe for utility application.

Propulsion

Although substantial programs involving reactor mobility have indicated the possiblility of nuclear-fission-powered aircraft (ANP), nuclear rocket (NERVA), and reactor in space (SNAP-10A), mobility of nuclear fission power has been demonstrated most practical for ship propulsion, the United States, the United Kingdom, France, and the USSR have nuclear-propelled naval vessels. Commercial nuclear-powered ships have been built by the United States (Savannah—22,000 SHP); Germany (Otto Hahn—11,000 SHP); USSR (Lenin—2 × 90 MWt); and Japan (Mutsu) and in general have been technically successful (except for Japan's shielding problems) but uneconomic. At present no nuclear-propelled commercial ship is under construction or committed.

Electric Utility

The nonmobile use for nuclear energy has been to produce electric power for electric utilities. Table 19-4 lists nuclear plants of the world by country. When the world total of nuclear power (405,768 MWe) is operating at a capacity factor of 65 percent, it will provide energy that would require 10.5 million barrels/day of oil to replace. U.S. total reactors planned (182,015 MWe), when operating at 65-percent capacity factor, will provide energy that would require 4.7 million barrels/day of oil. Thus worldwide nuclear-fission energy will supply 16 percent as much energy as oil does in the world today, because total world consumption of crude oil in 1980 is about 65 million barrels/day. Domestically, nuclear will supply 30 percent as much energy as oil does in the United States today. U.S. oil use in 1980 averages about 15.9 million barrels/day, of which utilities use about 1.5 million. Utilities have been told by the government to reduce oil consumption 0.75 million barrels/day by 1990. U.S. nuclear-power output is about equal to that from the rest of the world.

Figure 19-3 shows the buildup of nuclear power and the oil displaced in the United States, France, and Japan. U.S. totals were higher, but the U.S. nuclear program has had many plants delayed and cancelled.

Cogeneration

Other than for electric power, only one major reactor is under construction in the United States. The Consumers Power-Dow Chemical cogeneration plant at Midland, Michigan, consists of two nominal 800-MWe PWR units. One unit will produce 800 MWe; the other unit will produce 500 MWe, plus 300 MWe effective power as steam: 0.4 million pounds per hour of 557°F-900 psia steam directly from the steam generator, and 3.6 million pounds per hour of steam from the high-pressure steam-turbine exhaust. It was necessary to insert an additional steam generator to provide a second

Table 19-4
World List of Nuclear Power Plants Operable, under Construction, or on Order (30 MWe and over) as of December 1979

Country	Total	Operable	Percentage of 1978 Total Electrical Capacity (%)	Total	Operable	Per Capita Nuclear Capacity (Watts)
1. Argentina	1,635	335	16	3	1	60
2. Austria	692		6	1		10
3. Belgium	5,450	1,650	53	7	3	560
4. Brazil	3,116		14	3		30
5. Bulgaria	1,760	880	25	4	2	200
6. Canada	15,356	5,476	22	24	10	670
7. Czechoslovakia	3,630	550	24	9	2	240
8. Egypt	622		15	1		15
9. Finland	2,160	1,080	24	4	2	450
10. France	45,768	8,238	87	53	16	850
11. East Germany	2,710	1,390	16	7	4	170
12. West Germany	26,568	8,000	34	28	11	420
13. Hungary	1,760		34	4		170
14. India	1,684	604	7	8	3	2
15. Iraq	900		?	1		80
16. Italy	5,295	1,387	12	9	4	90
17. Japan	20,436	14,522	17	30	22	180
18. South Korea	5,480	564	86	7	1	140
19. Libya	300		37	1		120
20. Luxembourg	1,250		108	1		3,500
21. Mexico	1,308		10	2		20
22. Netherlands	493	493	3	2	2	35
23. Pakistan	125	125	6	1	1	2
24. Philippines	1,240		34	2		25

Table 19-4 *(continued)*

Country	Total	Operable	Percentage of 1978 Total Electrical Capacity (%)	Total	Operable	Per Capita Nuclear Capacity (Watts)
25. Poland	880		4	2		25
26. Rumania	1,040		8	2		50
27. South Africa	1,844		11	2		65
28. Spain	13,322	1,073	49	16	3	350
29. Sweden	9,410	3,700	38	12	6	1,120
30. Switzerland	4,947	1,940	40	7	4	760
31. Taiwan	4,924	1,208	?	6	2	290
32. Turkey	440		9	1		10
33. United Kingdom	11,780	8,080	16	39	33	210
34. United States	182,015	51,169	32	189	71	830
35. USSR	24,795	11,915	11	41	25	100
36. Yugoslavia	615		6	1		30
World total	405,768	121,611		530	227	

Source: American Nuclear Society, Inc., "Nuclear News" (February 1980). Used with permission.

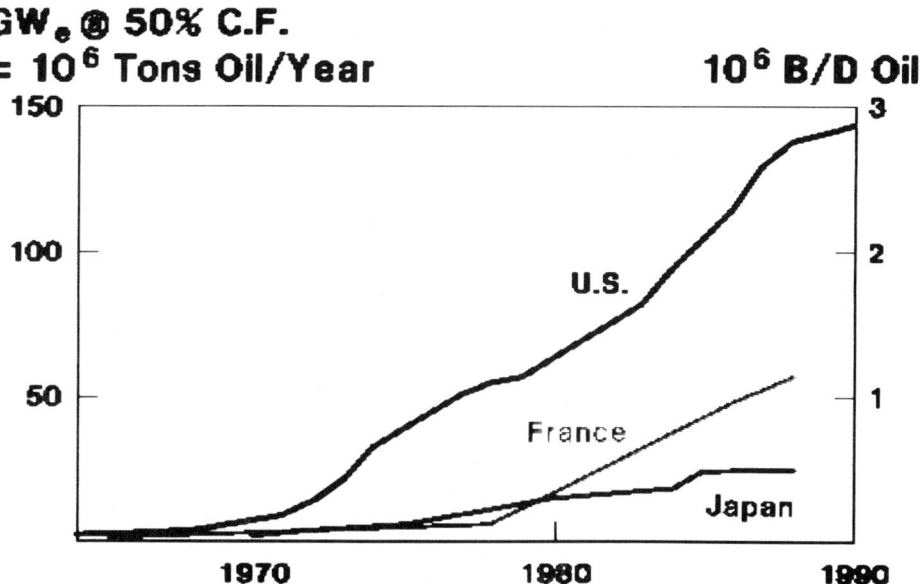

Source: From *Activites Scientifiques et Techniques,* Commissariat a l'Energie Atomique, Paris, 1976. Used with permission.

Figure 19-3. Nuclear Programs

barrier between radioactive reactor-coolant water and process steam because drugs for human consumption might be processed.

Desalting

Desalting seawater to provide fresh water is a large enough energy requirement to utilize the output of a large nuclear plant, in fact the Russian BN 350 (1,000 MWt) produces 150 MWe plus 2×10^9 BTU of heat for desalting water from the Caspian Sea. Worldwide, 1,000 plants are now turning saltwater or wastewater into clean water at a rate of 7 billion gallons per day. Water reactors have been analyzed and are technically adequate, but economics depend on the value of electricity and water at the particular site. A reactor that produces electricity for 50 mills/kwh could produce desalted water for an energy cost of $4/1,000 gallons.

Approximately 1,000 BTU/pound is required to produce steam, but multistage flash evaporators use heat energy more efficiently, so that a thirty-stage unit can produce steam at about 100 BTU/pound. A 1,000-MWe effective nuclear reactor could thus produce 100 million pounds of steam per hour and thus could irrigate an area of 500 square miles with a foot of water per year.

Enhanced Oil Recovery

Another potential application of fission energy that replaces oil usage and is large enough to use a nuclear reactor is steam injection for enhanced oil recovery. For heavy oils, about 1.5×10^6 BTU are required to heat injection steam for each barrel of oil recovered. In practice this means that one barrel out of each four recovered is burned to heat injection steam. A 1,000-MWe-equivalent water reactor would be adequate for recovering 160,000 barrels/day of heavy oil, which at $30/barrel is worth $5 million/day. To provide an economic thirty-year life for the reactor, operating at 65-percent capacity factor, would mean recovery of 1.1 billion barrels of oil. If recovery amounted to 20 percent of the oil in the ground, the oil reservoir would have to contain 5.5 billion barrels originally, within piping distance from the reactor. For example, the Orinoco heavy-oil belt, with 2 trillion barrels in a 400-mile × 40-mile area, averages 5.5 billion barrels within a radius of 3.7 miles. Economic reservoir requirements could be decreased to 1 billion barrels if the reactor were reduced in size to 600 MWe equivalent and cogenerated 400 MWe of electricity in addition to 2×10^9 BTU/hour for oil recovery. This would mean a maximum piping distance for steam of only 1.6 miles in the Orinoco oil belt.

Although steam is usually injected in wells less than 2,000 feet deep, PWR 1,000-psi steam could be compressed to 3,000 psi, which would overcome hydrostatic pressure to a depth of 6,000 feet, the maximum depth for Orinoco heavy oil. A steam compressor the size of a garbage can could compress 1 million pounds of steam per hour.

As oil prices increase, coal and nuclear energy become increasingly attractive for heating steam to inject in wells for enhanced oil recovery. Coal or nuclear energy provides lower cost BTUs for steam heat and also make more oil available for sale because oil is not burned for heating steam. The profit margin per barrel of oil sold might be twice as high with nuclear or coal heat, compared with oil heat, depending on how prices are escalated. Coal is already planned to replace oil for heating steam in some enhanced oil-recovery programs, for example, the $6 billion Shell Oil Company Alsands project in Canada. It may be that heavy-oil refinery bottoms will provide coke for some of the heat energy necessary for steam injection.

Uranium Enrichment

Energy Requirements

A large energy requirement associated with global fission energy is the uranium-enrichment process. Uranium enrichment by gaseous diffusion

20 Nuclear Energy—The Next Phase: Approaches to Restoration of Confidence and Motivation

Peter Fortescue

Introduction

This chapter is concerned with the notion that a successful and timely emergence from the present hiatus in the development of nuclear energy in the United States may well depend on the degree to which the lessons of recent events will be incorporated in the design, operation, and regulation of future plants. The present pause could furthermore provide the only available opportunity for the reassessment that may be demanded. Two areas particularly affected are the general safety philosophy and the criteria applied in choice of reactor systems.

Reactor safety so far has centered almost exclusively around satisfaction of the letter of regulations aimed at physical protection of the public. As a result, little regard has been paid to other consequences of malfunctions otherwise safeguarded by the measures taken. What is thus overlooked is that public *confidence* in declared safety is lacking and is undermined at present not by physical harm experienced but by the sense of uncertainty conveyed by dramatically presented reports of breakdowns, all too often involving protracted anxiety and obvious financial burden to consumers and plant owners alike.

Thus, simply following regulations is not enough. Also needed is a fuller appreciation of the merits of systems that are inherently tolerant of fault and are readily reparable, so that these criteria may receive their due regard in design selection.

As for the other area mentioned, the relative attraction of competing reactor systems has up to now been assessed primarily on the basis of near-term economic projections, which inevitably tend to favor progress by modification and extension of developments already undertaken. Aside from bypassing the question of reliability, these estimates have greatly outweighed considerations of nuclear-fuel-resource conservation. This is of vital importance because a limitation to temporary usage would greatly diminish justification for the necessarily long-term aftermath that any nuclear employment would carry. This same theme, expressed in terms of transient need, is indeed also behind the opposition to nuclear power voiced by believers in the early availability of alternatives. Unwarranted faith in these alternatives, rather than ignorance of nuclear matters, here appears the most significant impediment to balanced public judgment.

Two happenings in the outside world—namely, the oil embargo and the Three Mile Island debacle—have occurred, however, which profoundly affect planning for nuclear power. The former by its dramatic exposure of the limitations of our resources, and the latter by the resistance it has created to what remains the most promising hope for their extension, together would seem to present a particularly unfortunate conjunction of events.

However, although the demand for energy can only grow, the prospect remains of salvaging the situation by modification of our approach to nuclear-power generation. The present doldrums at least afford some respite for necessary reassessment. This chapter is concerned more specifically with some thoughts for consideration among the technical issues involved in such an endeavor.

Forgivingness

In addressing the subject of enhancing plant reliability, it should first be made abundantly clear that the ultimate safety of plants existing, or now under construction, is not in question here. The objective is simply to explore means to avoid the continued recurrence of headline-garnering incidents, the frequency of which even now threatens access to an essential resource and will inevitably increase in proportion to the number of reactors built unless something is done about it. This last point is of great importance because, thanks to modern communications, an event anywhere is an event everywhere. The bare achievement of satisfaction for the 150-GW system now under various phases of completion here would have to be improved some twentyfold to provide a similar level of alarms in a world possibly employing some 3,000 GW in thirty or forty years.

The key question is whether the required improvement appears economically attainable by improvements in the design of the present leading system (the light-water reactor, or LWR), or whether one of its present competitors offers a substantially better prospect in this particular area, or even whether some quite new start is indicated. Since the first of these possibilities is no doubt already under intensive study, and the third represents a last resort, it seems appropriate here to begin with a closer look at the second alternative and, more specifically, to examine claims of the high-temperature gas-cooled reactor (HTGR) system in the areas in question. These claims particularly well illustrate the key issues concerned.

Because the total impact of a considerable number of separate items is involved here, some ordering is necessary in the interest of clarity. It is convenient first to consider those differences between water and gas cooling that can properly be regarded as basic, in that they apply regardless of particular design. Next there are factors specific to the use of helium in association with an all-ceramic quasi-homogeneous core, as typified by the HTGR in the United States and the Thorium High/Temperature Reactor (THTR) in Germany. Finally, there remain factors that offer further advantage when exploited by appropriate design of this class of reactor.

Fundamental Factors

The most fundamental property of any gas that is noncondensable over the whole range of temperatures encountered is that it always totally occupies the space it is in and, so confined, obeys one unique and linear temperature-pressure relationship. Because there is no liquid-gas interface to be considered, a single unambiguous signal (pressure) always suffices to determine just where the coolant is. This, most importantly, implies that the most rapid depressurization can be accommodated without any concern for the several possibly dire consequences of void formation, including local core dryout and cavitation of pumps, as would be the case with liquids confined at temperatures above their atmospheric boiling points. Elimination of the necessity to await the possibly protracted cooling of water systems before atmospheric pressure can be reached is an advantage recently made evident at the Three Mile Island plant. The working pressure required by gas-cooled reactors is furthermore typically much lower than that needed to pressurize a water coolant adequately. This feature and others later mentioned indeed enable sufficient cooling protection to be readily provided for the HTGR class of reactor even at atmospheric pressure. In this sense it can be claimed that they are immune to coolant loss. These features, it is believed, substantially relieve dependence on operational judgment, in addition to extending the scope of remedial actions.

HTGR Application

This class of gas-cooled reactors is principally characterized by an absence of metallic cladding or core structural material and by the employment of an effectively homogeneously dispersed fuel. These features, only possible with an inert gas coolant, in fact considerably extend the scope of the influence of gas cooling on safety-related matters. The most important elements involved here are the relatively low power density and correspondingly high core thermal inertia and the high allowable structural temperatures associated with this type of design.

The overall power density of the HTGR core is, in fact, typically more than an order of magnitude lower than that of an LWR, both the fuel and the moderator (graphite) being furthermore in sufficiently intimate contact to ensure an immediate contribution to heat capacity from the moderating material. This point also distinguishes the thermal behavior from that of a heterogeneous gas-cooled core and results in an extremely slow temperature response. Also in association with removal of the temperature limits normally set by metallic fuel cladding and internal core structure, invaluable time is provided in the event of need for remedial action. These features furthermore greatly facilitate the provision of adequately maintained cooling at atmospheric pressure, even following the most rapid credible depressurization from full-load conditions.

It is particularly important, moreover, to realize that thermal inertia buys much more than just time for action, for the delay greatly reduces the eventually necessary heat-removal rate. At the same time, the substitution of the very high temperature associated with graphite vaporization for that of metal melting as the prime limitation provides a specially high driving-temperature differential for the process.

In the case of the Fort St. Vrain reactor, the available temperature gradient can safely conduct the residual heat to the water-cooled prestressed-concrete reactor vessel (PCRV) liners, even in the absence of pressurization and a total cessation of coolant circulation. As a matter of fact, on several occasions the Fort St. Vrain plant has undergone a stoppage of forced circulation immediately following scram from a high power level, for periods of a quarter hour, and it has done so without appreciable graphite temperature increase or other cause for concern. This surely well illustrates what is meant by a "forgiving" design, for with any other reactor type such a sequence would certainly have ranked high among causes for anxiety and substantial service interruption.

The general Fort St. Vrain experience also, unfortunately, illustrated that forgivingness, though a highly prized attribute, is not synonymous with reliability; for this plant has so far not been notable in the latter respect. It can be said, however, that auxiliary systems rather than the reactor were the culprit here—a familiar situation and recognized as such by the expressed views of its owners.

The use of a PCRV for integral containment of the entire primary system is a feature of the HTGR that is rendered practical only by the use of a noncondensable coolant; therefore, benefits accruing from this may be properly counted among the attributes of gas cooling.

As far as safety features are concerned, the attractions of PCRV containment arise primarily from the redundancy of the tension-bearing members (cables), the barrier to fault propagation that their independence provides, and their natural shielding from the effects of irradiation. The separation of the functions of sealing and load carrying represented by the use of a steel liner always in compression for the former duty further limits the possibilities of fault propagation. The total enclosure provided extends this feature to the entire primary circuit. The necessary liner-cooling arrangements moreover furnish an additional available heat sink.

As with other reactor types, independent auxiliary cooling systems and a hermetic secondary containment are of course also envisioned. Other safety-related features of helium as a coolant stem from its chemical inertness. Aside from facilitation of repair and maintenance associated with the absence of active corrosion products, this property also avoids the exothermic reactions possible between water and Zircaloy cladding, and the hydrogen production associated therewith. The fact that helium will not burn in any environment and latent heat is not involved reduces the effective energy content to that of its pressurization alone, which is relatively very small.

Another whole area of vital concern with respect to both safety and maintenance relates to the possible migration of fission products. Here the distinctive feature of the HTGR type of core resides in the additional barrier presented by its use of coated fuel particles. These, in effect, constitute tiny, independent pressure vessels that, by virtue of their small size, have the strength easily to contain the highest fission-gas pressure that can be generated. The use of trapping for continuous cleanup of the primary circuit further reduces the residual activity that may be present.

The prime reason for singling out the HTGR system for this extended comment is not mere advocacy of a particular reactor design but, rather, that this case so well illustrates the kinds of factors that need recognition in the reassessments proposed. This example also serves well to draw a distinction between *forgivingness*, meaning benign reaction to the perturbations of unplanned events, and *reliability*, which is concerned with reduction in the incidence of such happenings and with the time required fully to recover from them. Thus defined, forgivingness may be seen as nevertheless contributing most significantly to reliability by limit of damage and facility of repair.

Nuclear Fuel Resource Conservation

The other item noted as a factor misguiding progress concerned current neglect of the limits of fuel supply even in the nuclear area. The prime issue here is, of course, related to the need for some kind of assistance from breeder reactors, which is indicated even by combining the most optimistic resource figures with the most pessimistic view of the hopes of the developing nations.

Although it is not the purpose here to consider changes in the direction of breeder development, which is a less urgent matter, it is nevertheless important that the earlier thermal-reactor operation should not preempt the breeding option by ineffective prior fuel usage. Furthermore, any doubt about subsequent substantial breeder deployment must thus surely intensify rather than reduce interest in nuclear-fuel conservation. This matter is of immediate concern because, unfortunately, near- and long-term ends conflict.

If we simply want to get the most out of existing fuel with the least trouble, and if we regard fissile material appearing in spent fuel only as a serious nuisance and even as a menace, then the present use of light-water reactors represents a good way to go about it, and development should be along the lines of securing higher internal burn-up. Such a course will inevitably be indicated by looking at immediate cost figures alone. However, since the present operations produce only about one plutonium atom for posterity for every five natural fissile atoms destroyed in the process (and increased burn-up makes this exchange even worse), this represents a poor husbanding of a fissile material essential to subsequent breeding operations.

Another factor to be considered is the possible use of the thorium cycle. The importance of this goes far beyond mere provision of another fertile material resource, for this cycle yields a substantially improved neutron economy in thermal reactors and, in so doing, can furnish a nuclear-power system that is self-sufficient in fissile material without need for a preponderance of breeders. This, moreover, renders practical the better security of such breeders that could be provided by their collocation with processing plants in suitable enclaves.

The only point intended here, however, is that unless such long-term considerations have their appropriate weight in our selection, we may bequeath an impoverished legacy.

General Conclusions

Reexamination of the rules and criteria that have guided nuclear-plant design and operation up to now, has been building up steadily as the result of unfolding experience and has now become urgent because of recent momentous events.

In looking broadly at causes for the many criticisms that the establishment of nuclear power has to face, it is evident that critics variously express a common lack of confidence in assurances. Although time may take care of unwarranted fears, we cannot afford to wait; furthermore, there remains a residue of justifiable objection requiring satisfaction.

It is particularly important now, therefore, to take the opportunity to reexamine our directions that the present pause allows. In such a study the basis for design selection, including its suitability to societal needs, is particularly important; this is what really molds the form of plant constructed and, together with the rules followed in its operation, ultimately determines the quality of the service supplied.

The first necessity, however, is to ensure the safe operation, at whatever cost, of those reactors demanded for continuity of supply in the interim. The continued use of nuclear energy may well depend on their record of satisfaction. Improved operating procedures are particularly important here since these can be pursued independently of plant modification.

The question of how a plant is operated involves both safety and economics, by way of the influence of actions to secure the former on the continuity of service provided. It is in this area that past outlooks have been most drastically changed by the advent of nuclear energy. It is no longer possible to count on the bending of rules to maintain operation.

It may well be that some residue of this tradition did indeed enter into the optimism of earlier nuclear-plant-load factor projections, and that the effect of the newly emerging discipline has contributed in large measure to the decline from early expectations. If so, it is particularly important to search for safety measures that least conflict with the continuity of service, which is perhaps the most powerful factor needed to restore the confidence of all, particularly of the plant investors.

The HTGR class of reactors might contribute importantly to this end, particularly by way of its tolerance of fault, the inherent nature of its safeguards, and the facility of its repair. The validity of these claims would certainly seem to warrant their close examination in any forthcoming general study. However, voices from high places—recently reported to hold that troubles with present reactor types should preclude rather than encourage examination of variants—do little to encourage hope in subsequent reasonable action.

21 How Available Is the Nuclear Option?

William H. Hannum

Introduction

Energy ministers and heads of governments of the major industrialized countries specify that we must make much greater use of nuclear energy by the end of this century. Developing countries give ample warning that their needs are just beginning to be felt. Experts are unanimous that the age of oil is finished and that coal and nuclear must be used to displace oil.

Yet the facts today point in a different direction. It is hardly news anymore that current nuclear projections are substantially less even than last year's low estimate. The progressive diminution of the projections, as well as the effective moratorium on orders in most countries, lead many reasonable people to conclude that the nuclear option will never be more than a local solution in those few countries that have no feasible alternative.

What is the problem? Is more nuclear really needed? Is it really available? There is no technological factor that would preclude a much larger role for nuclear energy. The conclusion must be despite all the brave pronouncements, decision makers do not want nuclear. This chapter will consider some of the bases for this conclusion. First, it will deal with the reasons for concluding that there are no current technological impediments to nuclear energy.

Technological Areas

Area I: Resources

The most recent NEA report on uranium resources stated that there is likely to be an excess of available production capability throughout the next decade.[1] Current uranium production capacities now in place are based on the programs of five to ten years ago, when projections of nuclear deployment were much greater than they are now. By the mid- to late 1990s there will be increasing dependence on new capacity, utilizing resources beyond those currently classed as known. Thus the next fifteen or so years are considered a realistic lead time for exploration and development of production capacity. It will be necessary to maintain the momentum of current efforts, but uranium supply will not be a limit to nuclear deployment during either the short or the intermediate period. For the longer term, the Nuclear Energy Agency (NEA) and International Atomic Energy Agency (IAEA) have recently concluded that there probably exist two or three times the currently known uranium resources.[2] There is no basis for estimating how much of this resource is discoverable and economically recoverable; hence, no prudent planner will base energy strategies on such speculation. This does indicate, however, that there is some flexibility in any projected critical supply date.

Underlying all this is the proved technical feasibility of the breeder. The availability of a means of improving efficiency of resource utilization a hundredfold means that there is simply no question of ultimate resource availability.

In summary, there are no short- or medium-term global resource constraints on the nuclear option, provided there is realistic advance planning on uranium prospecting and resource development. Further, there is no long-term constraint provided the breeder is deployed on a realistic time frame.

Area II: Industrial Capacity for Fuel-Cycle Service

Enrichment capacity today is in excess supply. Supplementary capacity for future demand is in construction, and further conditional plans are in hand. There is no reason to consider that this ever need be a constraint.

Spent-fuel storage and reprocessing are currently well behind projected fuel-discharge rates. However, the difficulties in these areas are neither technical nor industrial, but political, in the sense that there needs to be clear political agreement about how to proceed. Manpower training and experience are constraints in some areas, including the manpower requirements for a competent regulatory infrastructure, but it would seem that this problem could be solved, given the incentives.

With respect to industrial capacity, one way to measure what can be done is to refer back to the projections of a few years ago. Although what have been referred to as missed opportunities can never be regained, the previously projected rate of buildup could be obtained again if desired. Further, from the French experience it can be seen that on the order of ten years is required to put in place an integrated productive capability. On either consideration, by the year 2000, two to three times currently projected capacities would appear to be available from an industrial point of view.

Area III: Science and Technology

The LWR is proved technology. Several other types of reactors (HTR, Candu, and so forth) are also proved technology with varied experience

How Available Is the Nuclear Option?

bases in industrial application. The LMFBR is also largely proved technology (some specific questions remain, such as optimum steam-generator design), with a limited but growing experience base. For the LWRs the experience base is sufficient that there are substantial moves toward standardized design in the United States, France, Japan, West Germany, the USSR, and Sweden.

Scientific data and many methods are confirmed by international exchange. A vast reserve of scientific expertise is available here that can readily be deployed in finding constructive solutions to perceived problems.

Area IV: Safety and Environmental Protection

Much work remains to be done in codification of safety practice. Further R&D can be expected to reduce uncertainties greatly. There will continue to be debates about the best way to provide any given degree of safety and about how safe is safe enough. But for the currently competing reactor types and their associated fuel cycles, it is known that there are no fundamental phenomena that have been overlooked, and realistic estimates can currently be made of the direct and indirect risks of nuclear power. In most situations nuclear is clearly superior to feasible alternatives. This is explicitly demonstrated in the United States, for example, by environmental-impact assessments. Nevertheless, the siting of nuclear facilities is a difficult political problem, even in comparison with the siting of other demonstrably more disturbing industrial installations. Although it is not clear how to address this very real problem constructively, this is not primarily a technological problem.

Area V: Wastes

Although there is continuing concern over the management of radioactive wastes, it is an established fact that, for all categories of nuclear wastes, safe interim management arrangements exist. Enough is already known to provide confidence that the remaining data base and procedures concerning the disposal of highly radioactive and/or long-lived materials can be put in place by the time they are needed. R&D efforts are currently accelerating rapidly. The Swedish KBS study demonstrated that even today a fully adequate scheme could be implemented if necessary.[3] The increased attention being given to this area is lessening the fears of some that this problem will simply be left for future generations to solve.

The amount of R&D still to be done and the effort required to overcome the effects of past practice in some countries should not be minimized. But the nature of most long-lived wastes is such that ample time is available to develop the necessary data base. Nuclear wastes do not seem to present a legitimate technical impediment to proceeding with deployment of nuclear energy, provided there is responsible attention to pursuing the relevant R&D and to the operational demands of current and past nuclear activities.

Area VI: Economics

Current economic assessments are unambiguous. Nuclear generation is dramatically cheaper than almost any other available means of power generation (hydroelectric power frequently even cheaper). In the long term, if the petroleum-availability data have any meaning and assuming some fuel-efficient adaptation such as the breeder, again the economic advantage of nuclear is self-evident. In the midrange, economic projections are obviously more difficult. In general (except for hydro), nuclear-fuel-cycle costs are much lower and capital costs higher than for other energy sources appropriate for electricity generation. With the high costs of money, nuclear is penalized prospectively; but high inflation rewards sunk costs retrospectively. Since even in prospective high-discount economics nuclear competes favorably, it is clear, as evidenced by experience, that the retrospective sunk-cost benefit will give operating nuclear plants in general quite favorable economics when they are in regular service.

Availability of capital funds is a problem, particularly for countries burdened with an oil debt. But any other choice involves explicitly mortgaging future generations with the specific penalties that inevitably follow from profligate spending on consumables.

Attitudes

If the technological, economic, industrial, and environmental aspects of nuclear energy are in such good shape, what about the nontechnical aspects. This section discusses four essentially negative attitudes; a fifth, that of France, might be characterized as, "I'm going as fast as is prudent," which hardly needs elaboration.

Attitude I: There Is A Better Way

For some countries (such as those with large untapped hydro resources), this may be true. For others this is merely a euphemism for, "I don't like nuclear." Almost everyone has other options; the real question is the cost.

Austria, which chose heavier reliance on coal and electricity from COMECON countries, is having second thoughts. Sweden calculated the cost in terms of standard of living and decided that it liked nuclear after all. For the United Kingdom, doing away with nuclear would lead to the loss of an industry and to lost energy-export opportunities for nuclear, coal, and oil. In a time of short supply one can forego a viable option without paying a price.

Coal is a commonly mentioned global option. However, all analyses point to the need for a growing global role for coal along with nuclear, so that there is really no need to struggle too long with a global decision between coal and nuclear. The answer is "both."

Conservation as a substitute for nuclear should be approached cautiously. Conservation through efficiency of energy use can go a long way—perhaps as much as a factor of two—in energy consumption for the more profligate countries. But conservation can never replace the constructive use of energy; only impoverishment can do that.

It has been noted that if it were not for the current continuing economic near-stagnation, the reduced growth in nuclear capacity would be causing economic problems. That assessment admits that economic growth is coupled with energy consumption however one assigns causality, and notes that as long as we are content with zero or negative economic growth, we can get by without accelerating nuclear.

Attitude II: The Ancillary Implications Are Too Great

There are three ancillary areas: safety, nuclear-weapons proliferation, and societal effects. For safety, the technology and administrative infrastructure are in place to ensure that the risks are not that great. Although there are clear and significant risks, as well as environmental costs and penalties, rational analyses show these to be generally preferable to those of the feasible options. The difficulties lie in ensuring discipline and motivation for the infrastructure and in constructively addressing irrational fears.

With respect to nuclear-weapons proliferation, the objective of the world's safeguard efforts, including those of the IAEA and the Non Proliferation Treaty (NPT), is to provide decoupling of the civilian and the military uses of nuclear energy without undue penalty to either civilian or military uses. International Fuel Cycle Evaluation (INFCE) strongly underscores that there is a coupling but that the coupling is weak. Further decoupling is feasible if desired but will come only as a result of political agreement. Unfortunately, there is a tendency in seeking to improve this decoupling to forget that civilian nuclear energy is at least as closely linked to national security through the energy-supply route.

The third ancillary area is societal. Nuclear is one of the necessary underpinnings for continuation of affluent industrial societies. Those who seek a fundamental change in our societies feel that this would occur if the nuclear movement could be stopped. On this point their logic is impeccable. It does not appear to me, however, that the change would be for the better.

Attitude III: I Can't Afford It

This is not a responsible argument in the highly industrialized countries. For smaller or less-developed countries the problem is more serious, and not only in the sense of the pervasive oil debt and competing investment needs. There are also serious costs in terms of industrial and regulatory infrastructures, grid sizes (until someone markets a decent small plant), and manpower. The IAEA helped greatly with training and aid, and a number of countries have entered the field based on supplied-warranted installations. That system is currently working well in the Soviet area, but there are strains in the system elsewhere.

Invention and innovation are going to be required if this objection is to be overcome. The current discussions on assurance of supply of fuel and fuel-cycle services will go a long way, if successful, but the overhead and initiation cost will also require attention.

Attitude IV: They Won't Let Me

Perhaps the only general comment that can be made in this area is that one should be very cautious of generalization. In some instances there have been delays to permit the codification of environmental principles (for example, on thermal discharges); others have been associated with developing legal principles (such as NEPA), others for technological studies (such as the Swedish KBS study). In various cases essentially extraneous matters have used the nuclear question as a vehicle of convenience (for example, the northern territories and aboriginal rights questions in Australia). In other cases, the questions are ones of sharing the economic costs and benefits (for example, the Japanese fishing question). As long as this very legitimate type of question has greater priority than energy supply, this will be a real impediment to nuclear.

This phrase is also popular rhetoric today, referring to diplomatic or fuel-cycle service pressures from the United States or the other uranium suppliers in the name of nonproliferation. The traditional response is expressed in terms of a broadening of the desire either for independent capability or for multiple (redundant) reliable sources of supply of fuel and fuel-cycle services.

Conclusions

There are no technical, economic, industrial, or environmental impediments to nuclear making a much larger contribution to the world's energy needs than is currently projected. Such an undertaking would not be simple, however, and even meeting current projections implies devoted effort in a number of areas.

But on the nontechnical side there are some powerful arguments:

"There is a better way": But they all appear to lead, at best, to economic stagnation and to even greater pressures on other resources.

"The implications are too great": This assumes that the no-nuclear route has easier ancillary problems.

"I can't afford it": On the global scale this would be a cry of despair; on a more local level it is a call to rethink traditional institutional (including regulatory) arrangements.

I am persuaded that the balance of needs clearly favors nuclear, but today that seems to be dependent on the political response to arguments such as these.

Notes

1. "Uranium Resources, Production and Demand," OECD-NEA (1979).
2. International Uranium Resources Evaluation Project (IUREP).
3. "Handling of Spent Nuclear Fuel and Final Storage of Vitrified High Level Reprocessing Waste," KBS Stockholm (1978).

22 Comparisons of the Health Effects of Energy Systems: An Assessment for France

Philippe Hubert

Introduction

Apart from technical and economic considerations, the development of an industrial technology has a number of side effects whose assessment is necessary for energy choices. Health effects are analyzed here from this point of view. The kind of synthetic comparative study of industrial risks presented here is quite recent—the first of its kind appeared in the 1970s. Since then, however, as such studies developed, both methodology and quantitative evaluations have improved.

The uniqueness of this chapter is that it is based on French data and is therefore characteristic of a country with no fossil-fuel resources and within the Western European economic environment. It also provides data on risks in solar electrical plants as well as in nuclear, oil, and coal technologies. The following topics are examined: the various types of health effects; the need to develop scenarios; and the way to deal with specific effects, occupational effects, and so on.

Health Effects and Side Effects

The choices made in the field of energy depend not only on technical viability but also largely on comparative costs of competing forms of energy. In addition to direct production costs, however, the development of an industrial technology involves a number of indirect effects. Economic calculations have been developed to describe such phenomena and, possibly, to incorporate them into collective decisions. Assessment of the health and ecological effects of the various energy chains can be regarded as fitting into this general context.

The investigations cover not only indirect health effects, but also, more generally, the negative secondary effects of energy production that economists call *external effects*. As a rule the comparison made is between

---
Extract from a CEPN study under a EDF-CEA contract with the participation of MM. J.F. Belhoste, B. Durand, F. Fagnani, C. Maccia, T. Meslin, and J.P. Moatti.

conventional and nuclear energy, but a similar approach can be used for a prospective analysis. This is the case of the solar-energy risk assessment made here, as well as of the new methodological problems we encountered. To start with, an outline of the areas covered in these studies (table 22-1) shows that the consequences in terms of public health represent only one aspect of the indirect effects under consideration. However, will be seen in the conclusion, other side effects must be kept in mind when interpreting the results of risk assessment.

Classification of Health Effects

The number and variety of effects on health and the natural environment of such complex technological methods are potentially infinite. The purpose of this summary is not to attempt an exhaustive survey, but to try to provide as accurate a picture as possible of the salient factors. A choice of this kind is not problem free, and how arbitrary it is depends on the information available. Quantified data are most likely to provide objective information, and these will of course take precedence. But although this is certainly the best way of presenting the results of a comparative approach of this kind, one must remember that it introduces a certain bias. As can be seen in connection with radiological risks, the availability of a large amount of quantitative information on certain aspects of a risk does not necessarily mean that this risk is of special importance in relative terms. It means only that the risk has been subject to extensive research and monitoring. This is also the case with occupation hazards, where the large amount of data collected by social-security and insurance companies could lead to overestimation of that risk.

All these risks can be:

**Table 22-1
External Effects of Technologies**

Consumption of nonrenewable resources
Effects on the natural environment
Local social and economic effects
Health effects
Social and political effects
Macroeconomic effects
Geopolitical effects

Health Effects of Energy Systems

1. common to industry, or specific to an energy chain;
2. due to accident or normal operation;
3. short-term or long-term effects;
4. public or occupational;
5. local or diffuse;
6. well known or not.

This is illustrated by table 22-2. One sees that pneumoconiosis endangers workers' health in normal operation and has short-term, quite well-known, localized effects. At the opposite end is carbon dioxide, which has a public, long-term, diffuse effect that is little known. The difficulty in drawing a line between normal and accidental risk appears here: for instance, the wreck of a tanker is "accidental," just as is a power-plant breakdown. But although the first occurs almost every year, the second has a probability of occurring only once in every ten thousand years.

Table 22-2
Risk Classification

Type of effect / Energy chain	Coal	Oil	Nuclear (PWR)
Normal operation			
Localized, short-term	Coal workers' *pneumoconiosis* Air pollution (power station)	Air pollution (power station) long-distance transfer	Short-lived radioactive effluent (public) Occupational exposure and contamination (uranium miners)
Diffuse, short term	Drainage water pollution (mines and ore preparation)	Hydrocarbon liquid releases	
Long term	CO^2 climate		Long-lived gaseous releases (reprocessing) Kr^{85}, C^{14}, H^3, I^{129}. Long-lived waste Genetic effects
Accident situations	Coal tip ⟨ fire / dam breakage	Crude-oil transport (tanker)	Nuclear safety (power station)
Other common risks	Heating up of waters, occupational risks, land taken up, liquid chemical waste releases, noise, and so forth		

Different Models for Different Risks

Assessment of health effects leads us to work in very different fields; figure 22-1 shows the links between the various evaluations. However, it should be noted that there is not one model for each risk, but a combination of models. The three main constraints have first to be linked together, for we must know the exact location of energy plants, their detailed design, and the precise releases they are allowed to produce.

There is, in particular, a direct connection between health effects and current regulations on health and safety. Clearly, the state and nature of the releases and waste, as well as the frequency and consequences of accidents, are closely related not only to the basic technology but also to the nature of these regulations. The latter may differ from country to counry and may change with time. It is therefore important, when carrying out prospective studies, to detail these regulations and their future development. Some of the U.S. studies in assessing discharges simply use the maximum concentrations defined by U.S. standards as the exposure values for populations, whereas in practice it is highly likely that the levels released will be lower than the authorized limits; this is the case in the nuclear industry. Details should also be given of safety measures that depend on the cooling process used, the types and source of fuel (as in the case of conventional chains), and the choice of site for the plant.

The choice of site, which will be discussed later, is very significant for the assessment of public-health effects. Where safety is concerned, as in the

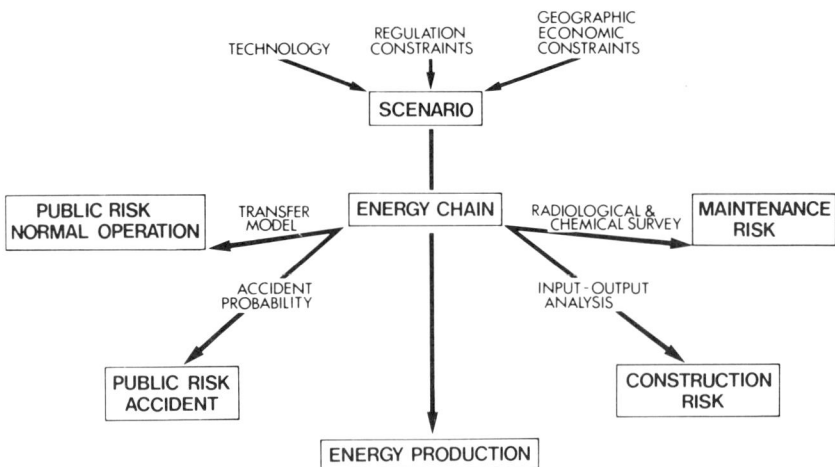

Figure 22-1. Risk-Assessment Pattern

case of health protection, the same techniques, when used in differing natural and human environments, will have very different health effects. Moreover, in addition to this evaluation in normal operation, we must assess effects of accidental situations by means of safety analysis.

Construction hazards must also be assessed, for the main risk can lie there. For that purpose we have to use an input-output macroeconomic analysis of quantify manpower requirements in order to apply occupational-risk ratios.

Developing Scenarios

The principle of comparison is simple: we count the effects related to the same electricity production. But therein lies the difficulty. Electricity is not like industrial goods, the main difference being that it cannot be stocked. The same amount of electrical energy has different values during peak and base periods. Thus the amount of energy production must be specified for a whole electricity-supply system.

That is why we have taken as a basis the French nuclear program as it was defined for the year 1990, the other energy chains being defined as the alternatives to this program. The energy supplied by this program in France by that time will be 296 TWh—that is, the basis of the loading curve. This allows us to define energy supply and installed capacity (table 22-3).

But power plants are only one part of an energy chain. The PWR fuel cycle is quite well known, but one should note that the French cycle is a closed one that includes fuel reprocessing. The oil chain includes extraction, transport, and refining (figure 22-2). The coal cycle also includes extraction and transport (figure 22-3). Another problem arises here: technological interdependence is not enough to define energy channels. A realistic approach is needed to determine what plants, transportation systems, and mines will be necessary to ensure the proper functioning of the program by 1990. The

Table 22-3
Scenarios: Level of Production in 1990, France

	PWR	Oil and Coal
Net production (TWh)	304.5[a]	296
Installed capacity (GWe)	63.5	52.2
Mean working time (hours)	5,200	5,670

[a]8.5 TWh are exported.

238   A Global View of Energy

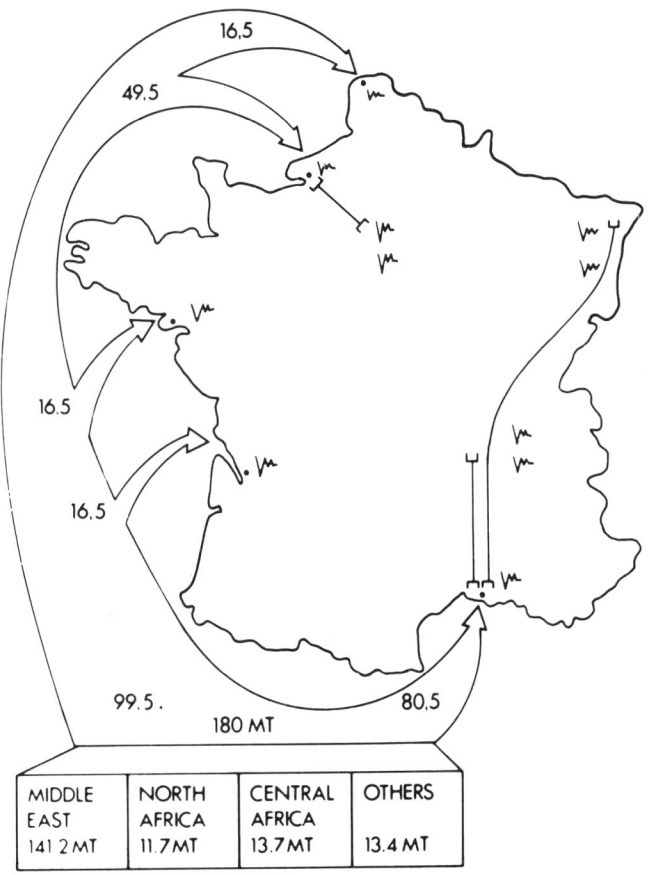

Source: Centre d'etude sur l'évaluation de la protection dans le domaine nucléaire (CEPN). "Risques sanitaires et écologiques de la production d'énergie électrique" (Paris, 1979).

Figure 22-2. Oil Scenario, 1990

amount of transported coal and the risks involved in extraction are at once geographical and technical parameters. The same applies to plant sites.

Developing a scenario is the only way to know where the plants should be located, where the fuel will come from, and so forth. There lies the specificity of the French case, for almost all fossil fuels have to be imported.

Health Effects of Energy Systems

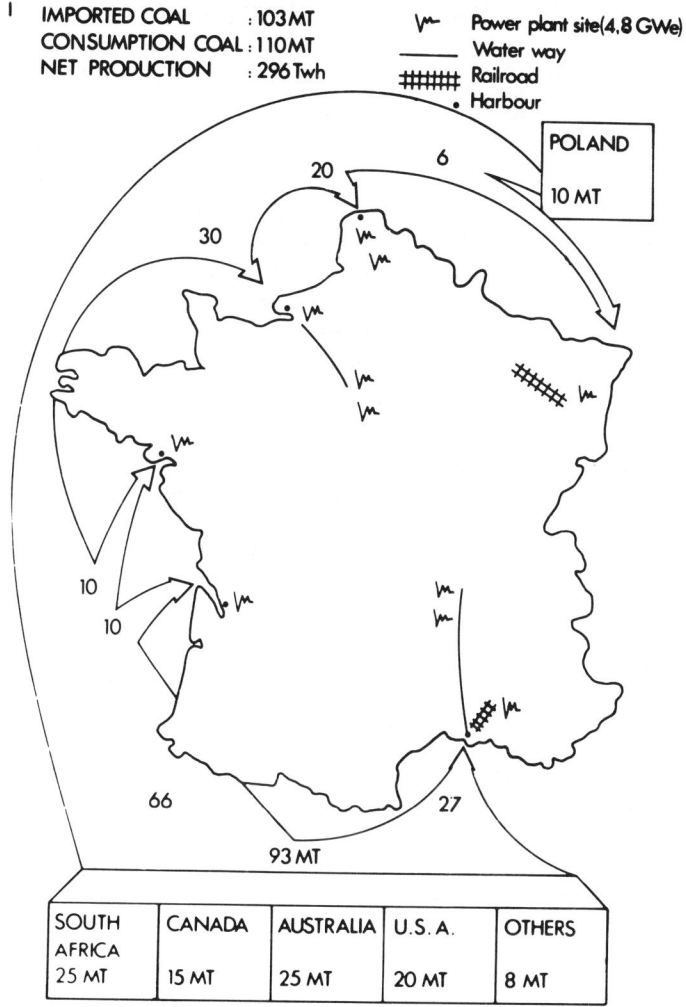

Source: CEPN. "Risques sanitaires et écologiques de la production d'énergie électrique" (Paris, 1979).

Figure 22-3. Coal Scenario, 1990

The origin of coal, for instance, is one of the most determinant parameters. The occupational risk due to extraction of the same amount of coal is ten times higher in underground mines than in open-air mines. The variability is about the same for oil extraction, but the total risk is much lower.

As almost any scenario can be prepared, we adopted a number of principles, the main one being that the energy channel to be set up is all that is necessary to provide 296 TWh in France by 1990, taking into account what exists at present and the origin of the fuel. (We should draw attention to the considerable imports due to fossil-fuel programs.)

Therefore, once the risk is reduced to an "average risk" for one energy unit, the term *average* refers in fact to the electricity demand in France in 1990, to the safety regulations, and to the countries from which we import fuel. It is in this context that we encounter the main methodological problems concerning solar electricity production. It is as yet impossible to build such a scenario for a solar system. If it is instituted, it will be marginal to a thermal system for at least half a century. Moreover, the concept of "same production" leads us to consider not only power-plant production, but also distribution and uses of electricity, as solar and classical systems will be very different at that point.

We hope we have made clear here that a potential energy chain is in itself a scenario and not only a technology. Almost negligible for the PWR chain, and low for oil, the difference is important for coal and determinant for solar systems.

Specific Risks

The two major specific risks are pollution and radiological effects. In both cases the assessment follows the same pattern; and these two risks are both assessed theoretically, in contrast to common risks, where knowledge comes from direct observation.

Briefly, the various phases involved in estimating the health risks of effluent releases are as follows (figure 22-4). First, primary releases have to be assessed (usually this involves not easily measurable elements, but estimates obtained from models based on a series of hypotheses); the variety of releases becomes very wide if a distinction is made between the various chemical or radioactive substances involved. Between the time of release and of human exposure, transfer in the environment takes place. Transfer models represent an important field in which a considerable amount of research has been conducted.

The problem varies according to the element under consideration—air or water—or the way in which transfer is effected: agricultural production, food chains, and so forth. The aim is an individual-exposure model based on the resident population, agricultural-production models, and consumption models. As the environment is not pollution-free nor is the radiation level equal to zero, the "zero point" of pollution and irradiation had to be evaluated.

The same model works in accident evaluation, applying it to the various accident sequences with their probabilities. We used the WASH-1400 report

Health Effects of Energy Systems

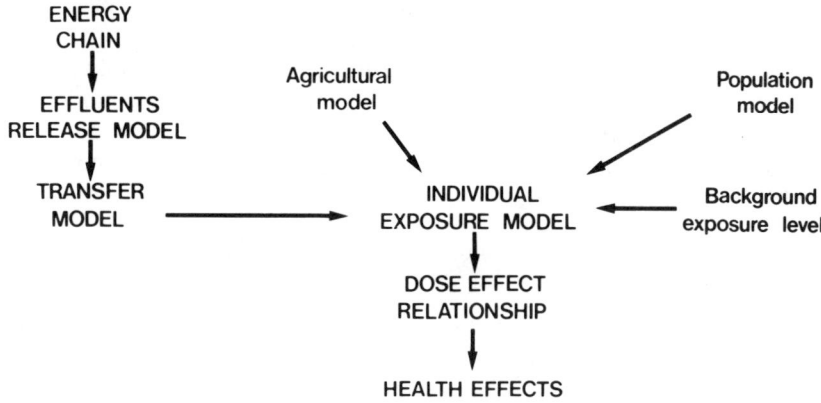

Source: CEPN. "Risques sanitaires et écologiques de la production d'énergie électrique" (Paris, 1979).

Figure 22-4. The Assessment of Health Effects

to do a crude adjustment to the French situation. The last steps of the model are used for assessing occupation effects. There we simply apply the ICRP relationship to the recorded exposure levels. Table 22-4 shows the yearly impact of the whole PWR scenario in France. The effects are hypothetical cancer fatalities. One should note that the effects are approximately the same for the public, here the French population (50 million inhabitants), as for workers (30,000 people involved). The other point is that the mathematical expectation of accidental effects is much lower than the impact in normal operation. However, one should be cautious, for safety studies are open to high uncertainty.

Such data are controversial in the case of atmospheric pollution since we know less about both exposure and dose-effect relationships than about

Table 22-4
Annual Radiological Risks; PWR Scenario (304.5 TWh in1990)

Normal operation, workers	
Total radiation-induced effects	7
Normal operation, public	
Total radiation-induced effects	5
Accident[a]	
Immediate death	0.0012
Short-term death	0.08
Somatic effect (long term)	0.026
Thyroid cancer	0.26

[a]Adapted from WASH-1400.

radiation. Using data from Lave and Seskin reviewed by Hamilton, we calculate the total yearly impact of an oil scenario for France to be about 200 deaths. This impact affects a population of about 25 million living less than 100 km from the sites. Despite all the uncertainties of the model, the absence of a desulfurization system in France, and the crudeness of mortality indicators, the difference obtained with nuclear power seems to be significant.

Occupational Risk

Workers are open at once to risks specific to the energy chain and to common industrial risks. The radiological risk has been already reviewed in a preceding section. The occupational nonradiological risk is due not only to the operation of the energy chain, but also, and mainly, to the construction of the facilities. Whereas the first is easy to observe, the second requires a special analysis, here a macroeconomic model. The total amount of investment needed for the construction is the input in a Leontief model, from which manpower requirement is computed. Risk is then assessed through accidental risk ratios.

The accidents have various consequences. An attempt has been made to aggregate them with a common indicator: Equivalent Working Day Lost (EWDL), defined as follows:

1 death	6,000 EWDL
1 occupational disease	338 EWDL

This allowed us to assess the results of figure 22-5, the risk due to the production of 1 TWh of electricity. The levels range from 1 to 10, oil being the safest and coal the most hazardous. The nuclear system has quite a low occupational risk, whereas the solar system seems to be relatively hazardous. This is because of the importance of manpower requirements (see figure 22-5). It is worth noting that construction represents the major portion of risk, with the exception of coal, where the contrary is true. This difference is due to the varying capital intensity in the chains.

There are two possible reasons for these differences in collective risk: either the manpower required is different or the individual risk is different. The corresponding values are shown in figure 22-5 and table 22-5. As far as the latter is concerned, coal is the most hazardous; but solar energy, PWR, and oil are on the same level. Figure 22-5 also shows the small importance of "specific" risk. This appears clearly for nuclear power, although the equivalence between radiation-induced deaths, which occur after a latent period of more than fifteen years, and immediate accidental deaths overestimates radiation-induced risk.

Health Effects of Energy Systems 243

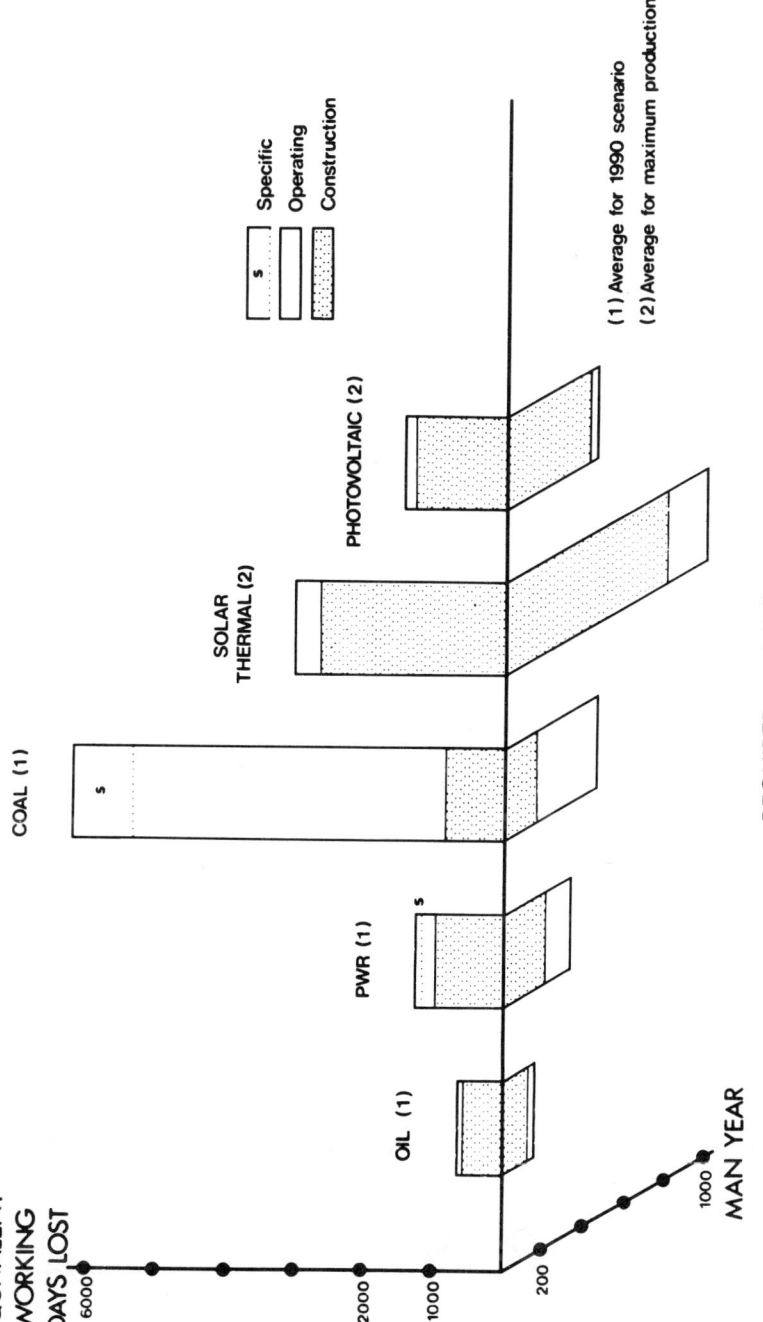

Source: CEPN. "Risques sanitaires et écologiques de la production d'énergie électrique" (Paris, 1979).

Figure 22-5. Collective Occupational Risk Normalized to a 1 TWH Production

Table 22-5
Individual Occupational Risk: Number of Deaths for 1,000 Man-Years

	Coal	Oil	PWR	Thermal Solar	Photovoltaic	Average, France
Common risk	0.50	0.16	0.15	0.13	0.16	0.14
Specific risk	0.12[a]	—[b]	0.07[c]	—[b]	—[b]	0.02[d]
Total	0.52	0.16	0.22	0.13	0.16	0.16

[a] Chemical risk: pneumoconiosis.
[b] No survey.
[c] Radiological risk.
[d] Officially recorded occupational diseases.

In the case of occupational risk, one should note the ambiguity of this type of analysis, which points to the fact that separating health effects from a global social assessment (including manpower and contribution to economic growth) seems irrelevant.

Synopsis of Effects

It is useful to give synthetic figures for all these types of risk. This requires a common quantitative indicator of effects. The one suggested for occupational risks, Equivalent Working Days Lost, fits quite well here. This is how table 22-6 was set up. We also used qualitative terms to describe the relative extent of the risk (+) and its uncertainty (?).

We need not stress how subjective the use of such indicators is; they are based on too many assumptions and value judgments. We suppose, for instance, that the life of a worker has the same value as the life of the man in the street. That is not true in actual decision making. One must also be cautious about quantitative evaluations that are sometimes based on fairly weak scientific evidence.

However, some conclusions can be drawn from this comparison. The nuclear chain appears to be relatively "clean" in normal operation. The effect on the public is much lower than that of air pollution from fossil fuels. Accident risk in the nuclear chain, insofar as data are reliable, also seems quite low. The coal chain, on the contrary, seems to be more hazardous than the other energy chains. Oil has quite low risk but still has important pollution effects.

Although it may not provide definite conclusions, such a synopsis is of great help in putting different evaluations of risk into perspective and, despite many uncertainties, does give a general picture of where the risk lies. The general perspective provided by table 22-6 is thus quite useful in

Table 22-6
Synopsis of Effects

Energy Chain	Health Effects on Workers[a]	Public-Health Effects[a]		Environmental Effects	
		Normal Operation	Accident Situation	Local	Overall
PWR	1,300	~100 (waste?)	~(10) (??) (power station)	—	??
Fuel	650	4,000 (??) (air pollution)	++ (??) (refinery)	+++(?) tanker accident air pollution water pollution	CO_2 climate
Coal	6,000 (miners)	2,000(??) (air pollution) (rail transport)	+(?) (tip)	++ ? air pollution water pollution	long-distance transport air pollution

[a]Equivalents in days lost because of accidents, death, and occupational diseases (collective-risk indicator).
[b]Population in the vicinity of plant (collective-risk indicator)

avoiding the pernicious effects of protective action. Reducing risk at any given point may give rise to some trade-offs: protecting the public can generate occupational risk; protective actions can be dangerous for safety; safety systems can be costly in terms of maintenance risks. There are also trade-offs between short-term and long-term effects, but little is as yet known on the subject. There is still much research to be done both in assessing the long-term risk itself and in developing a methodology to deal with it.

Conclusion

The aim of this analysis cannot be energy choices, for numerical values are too uncertain and physical constraints too strong. It is nonetheless a good basis for improving efficiency in risk management. It can also be considered an assessment of the social effects of an energy system, among other side effects. As far as the macroeconomic environment and determination of employment are concerned, developing scenarios and using an input-output model provided quite powerful tools for the broader analysis necessary to put health effects into perspective. It allows us to measure the social and economic impact of the energy changes that will take place in France before the twenty-first century.

References

1. Andurand, R. Le rapport de sûreté et son application dans l'industrie. *Annales des Mines* (July-August 1979).

2. Bliss, C., et al. Accidents and unscheduled events associated with non-nuclear energy resources and technology. EPA 600/7.77.016 (February, 1977).

3. CEPN. *Risques sanitaires et écologiques de la production d'énergie électrique.* Paris, 1979.

4. Farmer, F.R. Methodology of energy risk comparisons. *Colloque sur les risques des différentes énergies,* Paris, 1980.

5. Hamilton, L.D., et al. *European Scientific Seminar on Methods for Optimizing Protection in the Nuclear Industry.* New York: Pergamon Press, 1980.

6. Health and Safety Executive. *Canvey: An investigation of potential hazards from operations in the Canvey Island/Thurrock area.* London: HMSO, 1978.

7. Inhaber, H. Risk of energy production. AEC 1119, Canada (1978).

8. Organisation Mondiale de la Santé. Implications sanitaires de la production d'énergie nucléaire (1977).

9. Okrent, D. A general evaluation to risk-benefit for large technological systems and its application to nuclear power. UCLA-ENG-7777 (December 1977).

10. Reactor Safety Study (Rasmussen Report). WASH-1400, Nuclear Regulatory Commission, 1975.

11. Tanguy, P. Réflexions sur la comparaison des risques nucléaires et d'autres risques industriels. *Annales des Mines* (June 1980).

23 Health Hazards Associated with Electric-Power Production: A Comparative Study

W. Paskievici

Introduction

The risk-assessment analysis of various energy sources finds itself today at the intersection of four important streams of human activity: (1) a search for additional sources of energy, (2) a drive for improving occupational safety and general public health, (3) a movement directed toward protecting the environment, and (4) a call for more public participation in the decision-making process. Conflicting activities are sometimes associated with these different aims; it should not be surprising, therefore, to observe that risk-assessment analysis, a rational endeavor but one that operates in a social and political environment, is under public as well as scientific scrutiny and that the models, assumptions, methods, and results provided by such analysis are constantly challenged.

Risk analysis is usually justified by the assertion that it provides the decision makers with additional information that helps them choose the most appropriate combination of energy sources for future societal needs. In reality, however, the choice is determined by stronger forces than present concern with health hazards. Indeed, political decisions, such as maximal use of national resources or policies of full employment; economic reasons, such as industrial development or availability of energy at the minimum cost; and social considerations, such as public support all take precedence over risk assessments if, as it is now considered, the hazards associated with production of energy are not higher than hazards associated with other industrial activities and if the estimated benefits of energy outweight the costs associated with the production and the use of this energy.

Systematic efforts to quantify and compare risks of energy production have started with the advent of nuclear energy [1]. Further efforts have been devoted to including in the economic evaluation of energy alternatives the hidden social costs associated with health hazards and degradation of environment [2]. Despite many years of efforts (for example, [3]), there is no universally agreed-on method of carrying out comparative studies and, therefore, to rank energy sources according to their degree of safety [4, 5].

There are also acknowledged gaps in our understanding of health effects from chemical and radioactive substances released in the environment during the entire fuel cycle (from extraction of fuel to the disposal of wastes), and it is nearly impossible to agree on the economic value of human health and quality of the environment.

Risk analysis can be justified, however, on two other grounds. First, it can be a tool to identify areas in which more research is needed to improve our knowledge, such as the health effects of air pollution. Second, it can be a tool to identify activities within a fuel cycle that would require better control to diminish the health hazards. With these considerations in mind, a comprehensive study has been undertaken to review published papers on health hazards associated with the production of electricity from different energy sources [6]. This study attempts to do the following:

1. Establish the reliability of raw data.
2. Analyze the methodology used by the different authors.
3. Compare results.
4. Explain the reasons that these results are so different.
5. Find limitations of these studies.
6. Provide a "best" estimate of health risks.

This chapter will present some of the results obtained in that study. These results will focus on the critical review of published estimates. The next section briefly describes the methodology used in risk-assessment analysis and the scope of this analysis, and the two following sections are devoted to health hazards resulting from diseases and from accidents.

Methodology

A fuel cycle is defined by all activities needed to produce energy from that fuel. These activities are grouped into five major categories, as identified by the history of the fuel:

1. extraction;
2. processing;
3. transport;
4. conversion to electricity;
5. waste management.

Processing includes sizing and cleaning for coal; separation and cleaning for natural gas; refining for oil; and refining, conversion to UF_6, enrichment, fuel fabrication, and reprocessing for uranium. Transport and waste management may apply to several activities within the cycle.

Health hazards associated with construction and dismantling of facilities used along the cycle are sometimes included in the categories indicated here. Their contribution is, however, negligible unless all other health hazards are very small. Health hazards associated with acquisition of materials other than fuel have not been included, since these hazards are relatively small for conventional energy sources. (The contrary may be true for renewable energy sources [7].)

For each activity within a cycle, health hazards have been classified in two major categories: those resulting from diseases and those resulting from accidents. Within each of these categories, health hazards have been classified according to the severity of the hazard (fatalities, diseases, and/or injuries) and according to the victims (workers or the general population).

Finally, results have been normalized to the production of electrical energy equal to 1 gigawatt-year, which is equivalent to the yearly output of a 1,000-MW(e) electric power plant operating at 100-percent capacity. (1 gigawatt-year = 1 GW-y = 8.76×10^9 kWh.)

The studies which are reviewed in this paper are shown in table 23-1. Since there is no uniformity in the presentation of the results, a special effort has been made to present the data in a consistent way; some doubtful results have been identified by question marks. Some minor inconsistencies remain, but they have been discussed in the text.

Health Hazards Resulting from Diseases

Introduction

Health hazards resulting from diseases are very difficult to establish, and estimates strongly depend on assumptions made for their evaluation. In some cases uncertainties are so high that authors may hesitate to quantify these hazards. A review of available reports indicate several reasons for the differences in estimates of health hazards. First, there is the difficulty of making predictions about future occupational diseases, since, although working conditions are improving, present data reflect old conditions, and since large uncertainties in dose-effect relationship still exist. Second, there is the difficulty of evaluating long-term chronic effects of atmospheric pollutants resulting from fossil-fuel combustion. Direct methods of testing suspected pollutants for their health hazards are inconclusive, and epidemiological studies remain controversial. Finally, there is the difficulty of estimating health effects on future generations of uncontained toxic chemicals and long-lived radioactive elements originating during electricity-production activities. Depending on assumptions made about dispersion models, the temporal horizon, and the equivalence between health effects at

Table 23-1
Main References

Basic Studies	
WASH-1224, 1974	[2]
Hamilton, 1974	[8]
Hittman, 1974	[9]
Smith et al., 1975	[10]
Comar and Sagan, 1976	[11]
Transitional Studies	
OECD, 1976	[12]
MITRE, 1977	[13]
Morris, 1977	[14]
WHO, 1978	[15]
Inhaber, 1978	[7]
IEA, 1979	[16]
AMA, 1978	[17]
United Kingdom, 1978	[18]
Recent Studies	
Ramsay, 1978	[19]
Schurr et al, 1979	[20]
Morris et al., 1979	[21]
Inhaber, 1980	[22]
CEPN, 1979	[23]
CONAES, 1980	[24]

distinct historical periods, the results may be either negligible or overwhelming (in absolute terms).

The Coal Cycle

For the coal cycle occupational hazards are concentrated during coal extraction. The common disease of miners is pneumoconiosis, which may develop into fatal massive fibrosis. The present high rate of this disease is expected to drop as new standards are implemented. Estimates of this health hazard vary according to the assumptions made—the use of historical data as opposed to prediction of future trends. The figures presented here correspond to actual extraction conditions (for the United States, 50 percent surface, 50 percent underground mining).

Occupational Fatalities. Table 23-2 shows estimates of occupational fatalities from chronic diseases resulting from the entire cycle. These fatalities occur during extraction of coal. The highest values refer to old working conditions, the zero value to the assumption that radiological checkups would detect incipient pneumoconiosis and that the worker would

Table 23-2
Occupational Fatalities from Chronic Diseases, Coal Cycle

Author	Number
WASH-1224	0.008
Hamilton	0-4.7
Smith et al.	0-0.059
Comar and Sagan	0-4.7
Morris (1977)	0.1-8.7
Inhaber (1978)	0-3.5
AMA	0-6.4
United Kingdom	1-2
Ramsay	0-4.4
Schurr et al.	0-5.3
Morris et al. (1979)	0.11 (0-0.75)
Inhaber (1980)	0.11 (0-0.75)
CEPN	~0

then be removed from the mine. Since the most comprehensive study appears to be Morris et al. [21], the best estimate is taken to be 0.1 with an uncertainty factor of about 5 (in short 0.1(5)).

Occupational Diseases. Table 23-3 shows estimates of occupational chronic diseases; these diseases are essentially of respiratory origin (pneumoconiosis). The highest values correspond to historical data and include non-respiratory diseases. The lowest figures correspond to the most optimistic forecasts, based on a National Academy of Sciences (NAS) study [25] and on British data. The estimates of Morris et al. are based on data that refer to conditions before 1970 in the United States. The range in the French study is attributable to two different hypotheses of dust concentrations (2 and 4.5 mg/m^3); the additional small contribution, ~0.1, from other steps within the cycle, is due to occupational diseases such as asbestosis. It should be noted that the authors of several studies refrain from making predictions [19, 20, 24, 31]. Based on these considerations, the best estimate is taken to be 1.5(2).

Fatalities in the General Population. Table 23-4 shows estimates of fatalities in the general population from chronic diseases. These fatalities result from air pollution produced by burning coal and from ignition of coal-refuse banks. Since values shown in this table cover a wide range, a closer look is necessary. Figures that appear in the waste-management column originate from Hamilton [8], who makes a rough estimate of consequences of unintended combustion of coal-refuse banks. (On theoretical

Table 23-3
Chronic Diseases, Coal Cycle

| | Step | | | |
Author	Extraction	Processing	Conversion	Total
WASH-1224	0.8	—	—	0.8
Hamilton	13-60	—	—	13-60
Smith et al.	0.55-0.82	—	—	0.55-0.82
Comar and Sagan	0.8-64	—	—	0.8-64
Inhaber (1978)	0.55-0.82	—	—	0.55-0.82
AMA	0.8-64	—	—	0.8-64
Morris et al. (1979)	4.2-8.4	—	—	4.2-8.4
Inhaber (1980)	4.2-8.4	—	—	4.2-8.4
CEPN	0.69-4.74	0.06	0.03	0.78-4.83

grounds, such fires should be considered accidents.) With present standards, Holdren et al. consider that risks of such fires "should be nearly absent" [26].

With respect to risks associated with conversion, the situation is much more controversial. On the one hand, several studies [2, 24, 31] refrain from making numerical assessments by stating that the evidence is not conclusive. On the other hand, several other authors have made great efforts to provide such assessments. Their estimates vary according to assumptions made on SO_2 emission factors, on dose-effect relationships, and on affected areas.

Table 23-4
Fatalities in the General Population from Chronic Diseases, Coal Cycle

| | Step | | |
Author	Conversion	Waste Management[a]	Total
WASH-1224	?	—	?
Hamilton	4-133	?-13.3	4(?)-146.3
Smith et al.	0.2-36	0-13.0	0.2-49
Comar and Sagan	0.076-133	1.3-13.3	1.38-146.3
Morris (1977)	0-320	—	0-320
Inhaber (1978)	0.1-140	1.4-14	1.5-154
IEA	1.23-3.67	—	1.23-3.67
AMA	0.076-392	1.3-13.3	1.5-405.3
United Kingdom	?		?
Ramsay	0-32	—	0-32
Schurr et al.	0-10.7	—	0-10.7
Morris et al.	0-18.6	—	0-18.6
Inhaber (1980)	32-95	1.4-14	33.4-109
CEPN	0.34	—	0.34
CONAES	?	—	?

[a]Wastes originate from processing.

To illustrate these differences, let us consider four examples. With emission rates equal to present standards, Ramsay indicates a range from 0 to 32 arrived at by comparing results of three different models of dose-effect relationship: the Brookhaven National Laboratory (BNL) model (described by, among others, Morgan et al. [27] and extensively used by Hamilton and Morris) gives 32 fatalities for an average location. The NAS model [25] gives the range 6.7-20 (rural-urban site). A nonthreshold model, by Buehring et al. [28], gives the range 0-0.004 (rural-urban site). Schurr et al. [20] cut this range by a factor of three by assuming lower emission rates. The French estimate derives from the following assumptions: low-sulphur (1 percent) coal, no scrubbers, dose-effect relationships with threshold, and effects limited to 20 km. Finally, the Morris et al. [21] estimates are derived from a probabilistic calculation that gives, with 80-percent confidence limits, the interval 0-13.3 for fatalities associated with acid sulfates corresponding to present SO_2 emission limits, to which an interval 0-5.3 is added for fatalities associated with formation of polycyclic aromatics such as benzo(a)pyrene.

In conclusion, major uncertainties exist concerning health effects on populations resulting from air pollution from coal burning. Fatality rates attributed to these effects are between 0 and 20. (The BNL considers that long-range (>80 km) effects may be up to ten times higher than the short-range (< 80 km) effects considered here.) Within this range the best estimates are around 3, using a no-threshold linear dose-effect relationship, and around 0.2 using threshold dose-effects relationships. Our best estimate is 3(10).

Chronic Diseases in the General Population. Table 23-5 shows estimates of chronic diseases in the general population resulting from the coal cycle. These diseases are produced during conversation and include: (1) chronic respiratory-tract diseases in adults, (2) lower respiratory diseases in children, (3) related asthma attacks, and (4) aggravated heart-lung symptoms in old people.

Estimates may vary by a factor of 100 for the same author and quite considerably from one author to another. Such large variations can be explained by reasons given in the preceding section and by the fact that there are uncertainties related to diagnosis and reporting of diseases. The detailed analysis of the estimates presented here permits the drawing of the following conclusions:

1. Depending on the dose-effect model that is used, the number of estimated diseases can vary by a factor of 100.
2. Depending on the antipollution equipment that is postulated, the figures may vary by a factor of nearly 25.

Table 23-5
Chronic Diseases in the General Population, Coal Cycle

Author	Total
Hamilton	$(13\text{-}133) \times 10^3$
Smith et al.	$(1\text{-}216) \times 10^3$
Inhaver (1978)	$(1\text{-}216) \times 10^3$
IEA	$(5.3\text{-}15.4) \times 10^3$
Ramsay	$(1.17\text{-}117) \times 10^3$
Schurr et al.	$(0.34\text{-}34) \times 10^3$
Inhaber (1980)	$(190\text{-}570) \times 10^3$
CEPN	59

Note: It has been assumed that the average duration of an aggravation of heart-lung symptoms is seven days. All effects originate from conversion.

3. Depending on the location of the plant, the figures may vary by a factor of 10. As a best estimate, the value of 10^3 (20) has been taken.

The Oil Cycle

Since few if any oil-fired stations will be built in the future, the discussion of health hazards associated with the oil cycle will be very brief. There are practically no occupational health hazards, and for the population the hazards will be similar to those associated with the coal cycle, since in both cases the main toxic element is the same—sulphur and its components—and since SO_2 emission standards are similar. Table 23-6 shows estimates of fatal and nonfatal diseases for the population.

Table 23-6
Fatal and Nonfatal Diseases in the General Population, Oil Cycle

Author	Fatalities	Chronic Diseases
Hamilton	13-133.3	$(13\text{-}133) \times 10^3$
Smith et al.	0.04-36.5	$(0.2\text{-}218) \times 10^3$
Comar and Sagan	1.3-133.3	?
Inhaber (1978)	1.4-140	$(0.2\text{-}216) \times 10^3$
Inhaber (1980)	12-35	$(70\text{-}210) \times 10^3$
CEPN	0.31-0.62	111

Note: As for the coal cycle, we will consider that the number of fatalities and chronic diseases is around 2(10) and 1,000(20), respectively.

The Uranium Cycle

Estimates of health effects associated with activities within the uranium cycle are much more accurate and cover more areas than do estimates resulting from the coal or oil cycle. This is because of the specificity of radiation effects, the high degree of accuracy in measuring dose exposures, and the relatively accurate models that are used to calculate radioactive releases, dispersion of these releases through different pathways, and radiological effects. For these reasons most studies have reached similar results. The areas that are still in dispute are:

1. the "exact" dose-effect relationship (the discount method seems to be a good compromise between two extreme methods, one that neglects long-term effects by considering that for each generation the risk is negligible, and one that leads to very high absolute risks by integrating small effects on all future generations);
2. the degree of reduction in dose exposures obtainable by technological means;
3. the method of calculating collective dose commitments resulting from long-lived radioactive elements released along the fuel cycle, with particular reference to radon (generated by the thorium-230 family, of half-life $\sim 80,000$ years) released through uranium tailings, and carbon-14 (half-life $\sim 5,500$ years) released during normal operation of a nuclear power plant.

All references indicated in table 23-1 use no-threshold linear relationships with slopes in the range $1-2 \times 10^{-4}$ cancers/man-rem and in the range of $0.5-3 \times 10^{-4}$ serious genetic defects per man-rem. Reduction of dose exposures by engineered methods is essentially limited by cost-benefit analysis. Long-term (>500 years) effects either are ignored or are calculated by discounting future effects at some present worth (discount values of 2.5-5 percent).

The results shown here contain many data. It is neither possible nor necessary to review here all assumptions made by the different authors, since most estimates for total health effects are quite close, within a factor of ten.

Occupational Fatalities. Table 23-7 shows estimates of the number of occupational fatalities associated with activities within the entire cycle. These fatalities are essentially attributed to radiation-induced cancers. If one discards the very low values of WASH-1224[2], which underestimate dose exposures during reprocessing and conversion by optimistically assuming that early practices could be largely improved, the very high value taken by

Table 23-7
Occupational Fatalities from Diseases, Uranium Cycle

	Step					
Author	Extraction	Processing[a]	Transport	Conversion	Waste Management	Total
WASH-1224	0.013	0.007	—	0.04	—	0.06
Hamilton	0.13	0.05	—	?	—	0.18
Smith et al.	0.003-0.006	0.002	~0-0.004	0.03-0.11	0.074	0.11-0.20
Comar and Sagan	0.003-0.13	0.017-0.44	—	0.032	—	0.05-0.59
OECD	0.030	0.32	0.001	0.30		0.65
Morris (1977)	0.050	0-0.04	—	?		0.05-0.09
WHO	0.023	0.17	0.001	0.15		0.34
Inhaber (1978)	0.022-0.60	—	~0-0.004	0.034	0.074	0.13-0.72
AMA	0.003-0.13	0.017-0.44	—	0-0.13		0.020-0.70
Ramsay						0.09-0.18
Schurr et al.						0.05-0.27
Inhaber (1980)	0.022-0.60	—	~0-0.004	0.13	0.074	0.24-0.81
CEPN	0.035	0.024	0.001	0.150	—	0.210
CONAES	0.025	0.012	~0	0.124	—	0.160

[a]Processing includes refining, conversion to UF_6, enrichment, fuel fabrication, and reprocessing.

Health Hazards/Electric-Power Production   259

Comar and Sagan [11] for processing (the reason for such choice is unclear), and values that contain these figures, one readily comes to the conclusion that the postulated risk of fatal occupational diseases is situated around 0.20 (per GW-year) and that the most hazardous activity is conversion, which takes about 75 percent of the total risk.

Occupational Diseases. Occupational diseases refer to radiation-induced curable cancers and to nonradiological diseases. The number of diseases within the first category is generally taken to be equal to the number of radiation-induced fatalities. The number of nonradiological diseases appears to be about four times higher, as shown in table 23-8, derived from the French study. The major hazard seems to be silicosis among uranium miners. From this study, one can conclude that the total risk of occupational diseases is around one per GW-year of nuclear-generated electricity.

Fatalities in the General Population. Table 23-9 shows estimates of the number of fatalities resulting from cancers induced by exposure to radiation associated with activities within the entire nuclear fuel cycle. Two preliminary observations should be made. First, figures in the waste-management column are not comparable: for Smith et al. [10] and Inhaber [7] they represent health effects associated with wastes resulting from reprocessing; for Schurr et al. [20] they are estimates of health effects (discounted at 5 percent) resulting from long-lived radioelements found in uranium tailings; for Inhaber [22] they represent an upper-limit estimate of health effects resulting from releases from an underground waste-disposal storage. Second, in the extraction column the figures refer to health effects due to radon escape from uranium-mine tailings.

A detailed analysis of the figures shown in this table would indicate that the postulated risk of dying from radiation-induced cancers is situated, for the population, in the range 10^{-3}-10^{-2} per GW-y if one does not include reprocessing and the committed collective dose of radon. Reprocessing brings a contribution within the range 10^{-2}-10^{-1} per GW-y, but the effects of radon are strongly dependent on the assumed discount factor. To facilitate

Table 23-8
Occupational Diseases, Uranium Cycle

Diseases	Step				
	Extraction	Processing	Transport	Conversion	Total
Radiation-induced	0.020	0.024	0.001	0.150	~0.20
Others	0.68	0.038	0.013	0.031	0.77

Table 23-9
Fatalities in the General Population from Diseases, Uranium Cycle

Author	Step				Total
	Extraction	Processing	Transport	Conversion	Waste Management[a]	
---	---	---	---	---	---	---
WASH-1224[a]	—	—	—	0.021	—	0.021
Hamilton	—	—	—	0.067	—	0.067
Smith et al.[a]	—	—	<0.001	~0–0.013	~0–0.002	0.001–0.015
Comar and Sagan	—	—	—	0.013–0.21	—	0.013–0.21
OECD[a]	—	0.21	<0.001	0.015	—	0.22
Morris (1977)	—	—	—	0–0.21	—	0–0.21
WHO	—	0.068	<0.001	0.015	—	0.083
Inhaber (1978)	—	—	<0.001	0.03–0.23	~0–0.002	0.03–0.23
AMA[a]	—	—	—	0.01–0.21	—	0.01–0.21
Ramsay[a]						0.09–0.41
Schurr et al.[a]	0.031	←	0.001	0.03–0.23	0.005–0.25	0.09–0.41
Inhaber (1980)	<0.001	0.11	0.003	<0.001	0.25	0.074–0.53[b]
CEPN[a]						0.11
CONAES[a]	0.060	0.073	0.001	0.013	—	0.15

[a]See text.
[b]The total includes a contribution of 0.007–0.020 resulting from air pollution originated during fabrication of construction materials.

comparison with the results obtained for health effects resulting from the other fuel cycles, we will neglect the possible contribution of extremely long-term effects and select the figure of 0.1 fatalities per GW-y as the best estimate.

Chronic Diseases in the Population. The same conclusion holds true for nonfatal diseases in the population since the ratio of curable to noncurable radiation-induced cancers is about 1 to 1.

Genetic Effects. Many detailed calculations for the number of genetic effects have been made. Some of these take into account the genetically significant part of the total dose exposure (for example, radon is not genetically significant), as well as uncertainties related to dose-effect relationships. As a first approximation, it will be considered here that this number is equal to that of fatal or nonfatal cancers.

The Natural-Gas Cycle

For completeness, the natural-gas fuel cycle was also considered. According to available estimates, health hazards associated with the natural-gas cycle are negligible both for workers and for the public.

Summary and Discussion

The comparative study of health effects arising from different fuel cycles indicates that some major methodological problems in assessing these effects still exist, but that uncertainties associated with published estimates mainly reflect our limited knowledge in the physical, chemical, and biological reactions that relate emission of pollutants to physiological reactions.

As stated in the introduction, the main purpose of the original study [6] was to find the areas in which uncertainties are high and that therefore need more R&D, and to compare relative risks within the same cycle with a view to identifying the most hazardous activities. Notwithstanding the hazards of intercycle comparisons, the best estimates for each fuel cycle and for each category have been put together in table 23-10. The uncertainty factors should be considered not as real but as indicative of relative magnitude; for the uranium fuel cycle the range is also used as a graphical expression for indicating that the main uncertainty corresponding to the best value (the

Table 23-10
Comparison of Health Effects Associated with Different Fuel Cycles Normalized to 1 GW-Year of Electrical Energy (Accidents Excluded)

Fuel Cycle	Workers		Population		Total	
	Deaths	Diseases	Deaths	Diseases	Deaths	Diseases
Coal	0.1(5)	1.5(2)	3(10)	$10^3(20)$	3(10)	$10^3(20)$
Oil	~0	0.01	3(10)	$10^3(20)$	3(10)	$10^3(20)$
Natural gas	~0	~0	~0	~0	~0	~0
Uranium	0-0.2	0-0.2	0-0.1	0-0.1	0-0.3	0-0.3

Note: Best estimates are accompanied by uncertainty factors in parentheses.

higher limit) is associated with the shape of the dose-effect relationship and not with the computation of dose exposures for different activities within the fuel cycle. From this table the following general conclusions may be drawn:

1. The total risk for occupational disease is negligible for the oil and natural-gas cycles and similar for the coal and uranium cycles. For the coal cycle the major risks are located at extraction, whereas for the uranium cycle the major risks are located at the power station and at the reprocessing facilities.
2. The risk of fatalities in the general population is negligible for the natural-gas cycle, relatively low for the uranium cycle, and relatively high for the coal and oil cycles.
3. The risk of fatalities in the general population is lower than the risk of occupational fatalities for the uranium cycle, but it is higher for the coal and oil cycles.
4. The risk of occupational diseases is negligible for natural gas and increases for oil, uranium, and coal—in that order.
5. The total number of nonfatal diseases from coal and oil is much higher than from uranium; the natural-gas cycle shows practically no health effects.

Health Hazards Resulting from Accidents

Introduction

In principle, health hazards due to accidents should be easy to compute, since the cause and the severity of each accident are recorded as a matter of routine. Thus, from productivity figures and accident rates, occupational health hazards of each activity within the fuel cycle can be established. Corresponding figures for accidents affecting the general population can also be established from statistical data. It is only natural, therefore, to expect that estimates of health hazards resulting from accidents should be more accurate than estimates of health hazards resulting from diseases. In practice, however, several difficulties are encountered:

1. the lack of detailed statistics that specifically cover activities related to the energy fuel cycles;
2. the large dispersion in statistical data, which strongly discriminates against industrial activities with good safety records;
3. the difficulty of extrapolating from historical data, since trends depend on safety regulations, on productivity, and on the general economic climate;
4. the lack of uniformity in collecting data, which makes comparisons difficult from one country to another and even within the same country;
5. the determination of the degree of severity of an accident;
6. the computation of risks associated with very rare or hypothetical accidents.

This last difficulty deserves some attention since the public has an increasing tendency to focus on catastrophic accidents, either real or apprehended, even if the probability of such accidents is very small. For real but very rare events such as large explosions in coal mines, major fires of natural gas released from a pipeline, rupture of hydroelectric dams, and so forth, two difficulties are encountrered: the choice of the statistical method to determine the time of reoccurrence of such accidents if they are supposed to happen in a random fashion, and the evaluation of effects of technological changes on future accident rates.

For hypothetical accidents one can only rely on computational models to predict events in a frequency-consequence space. In doing so, three problems have been encountered:

1. the degree of confidence that can be placed in such calculations, as exemplified by the Reactor Safety Study [29];
2. the problem of equivalence between individual fatalities resulting from accidents of largely different consequences—that is, whether an event is more hazardous that results in a single fatality per year than one that would result in 100 fatalities every 100 years;
3. the fact that hypothetical accidents have not been considered in a systematic way for each activity and for each fuel cycle.

In spite of these difficulties, a comparison of health effects resulting from accidents has been made. Only estimates of real accidents will be compared here. As before, health hazards will be classified in four categories (fatal and nonfatal accidents, for workers and for the general population). Risks associated with construction or dismantling of installations have not been included, except where specifically indicated. As before, results have been normalized to the production of 1 gigawatt-year of electric energy.

The Coal Cycle

Occupational Fatalities. Table 23-11 shows estimates of occupational fatalities from accidents associated with the coal cycle. These accidents occur mainly during mining operations. A detailed analysis of these estimates would indicate a considerable agreement. Extreme values found in estimates by some authors, such as Hamilton [8], Smith et al. [10], Morris [4], IEA [16], Ramsay [19], and CONAES [24], generally correspond to different working conditions (for instance, surface or underground mining). For others, such as Comar and Sagan [11], and Inhaber [7, 22], they indicate the lower and higher estimates from other authors. The higher estimates for transport probably refer to total fatalities, since most of the authors do not differentiate between workers and nonworking victims. All single high values have been identified by a question mark and are not added to the totals. Finally, the most recent data confirm the decreasing trend for accident rates.

As a result of this analysis, the best estimate of the number of fatalities has been taken as 1.4, with a factor of uncertainty of 1.5, which is mostly related to the working conditions. The average risk seems to be lower in the United States than in the United Kingdom (1.8) or France (2.1).

Occupational Injuries. Table 23-12 shows estimates of occupational injuries from accidents associated with the coal cycle. The spread of values has been partly explained in the previous section. In addition, one should note that there are ambiguities in counting the number of injuries. Thus the French estimates should be considered not as indicative of very poor safety practices but rather as a different way of recording accidents. The French systematically include minor injuries in their estimates. As before, nearly 50 percent of injuries occur in mining activities. For injuries, the bast value has been taken as 60(1.5).

Risks Incurred by the General Population. Accidents associated with the coal cycle that affect the general population occur during the transport of coal, mainly by train, over the long distances that often separate the mines from the location of the power plants. Table 23-13 shows estimates of fatalities and injuries that occur during such transport. These estimates are derived from global statistics on rail accidents and are generally considered

Table 23-11
Occupational Fatalities from Accidents, Coal Cycle

| | Step | | | | |
Author	Extraction	Processing	Transport	Conversion	Total[a]
WASH-1224	1.28	0.027	0.073	0.040	1.42
Hamilton	0.4-0.8	0.05	—	0.013	0.46-0.86
Hittman	1.12	—	(5.0)?	0.09	1.21
Smith et al.	0.8-2.2	0.02-0.04	1.6-5.0	0.013-0.09	2.4-7.5
Comar and Sagan	0.60-1.30	0.027-0.054	0.073-0.53	0.013-0.040	0.72-2.0
MITRE	1.32	0.032	(3.06)?	0.016	1.37
Morris (1977)	0.27-1.47	0.05	—	0.013	0.33-1.53
Inhaber (1978)	0.7-1.5	←	1.6-5.0	0.013-0.090	2.3-6.6
IEA	0.135-1.35	0.05	(3.07)?	—	0.19-1.40
AMA	0.60-1.65	0.03-0.07	0.07-2.53	0.01-0.04	0.72-4.29
United Kingdom	1.4	←	0.2	0.2	1.8
Ramsay	0.4-1.3	0.03-0.05	0.04-0.11	0.01-0.04	0.48-1.50
Schurr et al.					0.40-1.33
Morris et al.	0.78	0.068	0.014-0.065	—	0.86-0.91
Inhaber (1980)	0.7-1.5	←	1.6-5.0	0.013-0.09	2.5-6.7
CEPN	1.90	0.08	0.12	0.024	2.12
CONAES	0.40-2.27	0.027	(3.07)?	0.013	0.44-2.31

[a]The total is not always equal to the sum because of roundings. It does not include values between parentheses.

too pessimistic. In particular, the high values of Hittman [9], MITRE [13], and IEA [16] derive from considering a ratio of 10 to 1 between injuries and fatalities. Although this ratio is valid for passengers, it is not valid for persons who are struck by trains. Therefore, in our judgment one should choose as best values 1.0(1.5) for fatalities and 1.8(2) for injuries.

The Oil Cycle

It is generally accepted that the risk of accidents is smaller for the oil cycle than for the coal cycle. There are several reasons for this: the activities within the cycle are more automated, the productivity is higher, the working conditions are safer, and the public is better protected. The activities for which concern persists are the extraction step (in the case of the workers) and the consequences of fires and explosions in refineries (in the case of the general population). Explosions of catastrophic consequences—100 to 1,000 victims—may occur, but because the probability of such events is very difficult to estimate, the risks associated with such hypothetical accidents have been ignored.

Table 23-12
Occupational Fatalities from Accidents, Coal Cycle

Author	Extraction	Processing	Transport	Conversion	Total
WASH-1224	53.1	0.93	6.8	1.6	62.4
Hamilton	17.3-40.0	4.0	—	1.6	22.9-45.6
Hittman	50.4	—	(48)?	8.5	58.9
Smith et al.	40-110	1.4-2.7	13-48	1.6-8.5	56-169
Comar and Sagan	29-65	3.5-4.0	0.4-31	1.2-2.0	~34-102
MITRE	65.9	3.4	(31.2)?	1.8	71.0
Inhaber (1978)	40-70	←	13-48	1.6-8.5	56-83
IEA	10-100	5.3	(31.2)?	—	15-105
AMA	29-106	3.5-4.1	0.4-31	1.2-2.0	34-143
Ramsay					70?
Schurr et al.					53-106?
Morris et al. (1979)	53	4.20	1-4.4	—	58.2-61.6
Inhaber (1980)	40-70	←	13-48	1.6-8.5	56-83
CEPN	1320	64.6	17.4	3.8	1406
CONAES					55-150

Occupational Fatalities. Table 23-14 shows estimates of occupational fatalities from accidents associated with the oil cycle. If one discards the extreme values shown in this table, namely the high value for processing, by Hamilton [3], and the low values for extraction by Hittman [9] and Smith et al. [10], the estimates of occupational fatalities are very consistent. As best choice, we have taken the value 0.35(1.5).

Occupational Injuries. Table 23-15 shows estimates for occupational injuries. Again, if extreme values are discarded, the results are reasonably consistent. Our best estimate is 30(1.5).

Risks Incurred by the General Population. These risks are considered negligible.

The Natural-Gas Cycle

There are few independent studies of accidents within the natural gas cycle; after the first studies, indicating that the risks were low, interest shifted to other sources of energy. However, the probable increasing use of liquefied natural gas (LNG) has spurred new studies in this area, in particular with respect to potential catastrophies. Estimates of consequences versus fre-

Table 23-13
Fatalities and Injuries from Accidents, Coal Cycle

Author	Fatalities	Injuries
WASH-1224	0.73	1.60
Hamilton	1.6	13
Hittman	5.0	48
Smith et al.	0.73	1.6
Comar and Sagan	0.73-1.73	—
MITRE	3.06	31.2
Morris (1977)	0-5.3	—
Inhaber (1978)	0.8-1.9	1.6
IEA	3.07	31.2
AMA	0.73-1.73	—
Ramsay	0.84-2.08	—
Morris et al. (1979)	0.40-1.76	0.82-3.68
Inhaber (1980)	0.8-1.9	1.6
CEPN	0.71	1.84
CONAES	0.67-3.06	—

quency of such risks exist [30, 31] but, as stated before, they will not be considered in this chapter where only estimates of real accidents are compared.

Occupation Fatalities. Table 23-16 shows estimates of occupational fatalities from accidents. If one discards the estimates by Smith et al. [10], which apply to synthetic gas obtained from coal, and takes into account the remarks made previously for the oil cycle, one could reasonably take 0.20(1.5) as the best estimate for the number of occupational fatalities.

Occupational Injuries. Table 23-17 shows estimates of occupational injuries. The same reasoning as before leads to the best value of 15(2).

Risks Incurred by the General Population. It is estimated that the risks from accidents for the general population are negligible.

The Uranium Cycle

The major concern in the uranium cycle is the possibility of accidents with radiological consequences. Very severe protective measures have prevented such accidents, and several estimates have shown that the potential risks—as measured by the product of the probability of such accidents and their consequences—are relatively small with respect to nonradiological

Table 23-14
Occupational Fatalities from Accidents, Oil Cycle

Author	Extraction	Processing	Transport	Conversion	Total[a]
WASH-1224	0.084	0.056	0.04	0.049	0.23
Hamilton	0.16	1.33	0.13	0.013	1.64
Hittman	0.01	0.04	0.01	0.05	0.11
Smith et al.	0.003	0.037-0.056	0.055-0.071	0.013-0.05	0.11-0.18
Comar and Sagan	0.08-0.28	0.053-1.33	0.04-0.13	0.013-0.05	0.19-1.73
MITRE	0.28	0.11	0.07	0.13	0.47
Inhaber (1978)	0.14-1.70	←	0.04-0.14	0.013-0.05	0.19-1.90
AMA					0.19-1.73
United Kingdom	0.3	←			0.3
Inhaber (1980)	0.14-1.70	←	0.04-0.14	0.013-0.05	0.30-2.0[b]
CEPN	0.12	<0.01	0.17	0.02	0.31
CONAES	0.26	0.106	0.06	0.013	0.47

[a]The total is not equal to the sum because of roundings.
[b]The total includes 0.09 fatalities which occur during construction.

risks. Since this chapter deals only with real accidents, risks from hypothetical accidents are not further discussed. (Risk assessments of nuclear reactor accidents are analyzed in Paskievici [6].) Therefore, the main hazards are occupational.

Occupational Fatalities. Table 23-18 shows estimates of occupational fatalities from accidents associated with the uranium cycle. The differences

Table 23-15
Occupational Injuries from Accidents, Oil Cycle

Author	Extraction	Processing	Transport	Conversion	Total[a]
WASH-1224	10.0	4.0	1.5	2.0	17.5
Hamilton	16.0	82.7	12.0	1.6	111.7
Hittman	0.58	3.15	0.58	4.75	9.1
Smith et al.	0.26	3.1-4.0	0.3-6.3	1.5-4.8	5.2-15.4
Comar and Sagan	10.0-28.0	4.0-82.7	1.5-12.0	0.80-2.0	16-125
MITRE	28.0	7.5	6.0	1.5	43.0
Inhaber (1978)	15-120		1.6-13	0.9-2.0	18-135
AMA					16-125
Inhaber (1980)	15-120		1.6-13	0.9-2.0	29-146[b]
CEPN	12.0	3.2	2.8	5.7	23.7
CONAES					43.0

[a]The total is not always equal to the sum because of roundings.
[b]The total includes 11 injuries that occur during construction.

Table 23-16
Occupational Fatalities from Accidents, Natural Gas Cycle

	Step				
Author	Extraction	Processing	Transport	Conversion	Total
WASH-1224	0.028	0.008	0.032	0.049	0.109
Hamilton	0.16	0.013	0.027	0.013	0.213
Hittman	0.025	—	0.004	0.05	0.08
Smith et al.	0.8-2.2	0.021-0.035	1.6-5.0	0.13	2.6-7.4
Comar and Sagan	0.028-0.28	0.008-0.013	0.027-0.032	0.013-0.049	0.076-0.37
MITRE	0.213	0.013	0.027	0.012	0.267
Inhaber (1978)	0.039-0.31	←	0.029-0.034	0.014-0.053	0.08-0.40[a]
AMA					0.08-0.37
Inhaber (1980)	0.039-0.31	←	0.029-0.034	0.014-0.053	0.16-0.48
CONAES	0.213	0.013	0.027	0.012	0.267

[a]The total includes 0.08 fatalities that occur during construction.

seen in these estimates are related essentially to the assumptions made by the different authors and only marginally to uncertainties of actual figures. The main factors that explain these differences are: (1) some authors consider mining as underground, but others consider 50 percent surface mining and 50 percent underground mining;(2) some authors include accidents occurring during the construction of nuclear plants, whereas others do not (unlike the case of the coal cycle, the difference is significant); and (3) some authors include estimated effects of accidents that may occur during the waste-disposal activities and during decommissioning of nuclear facilities.

A detailed analysis of the data shown in table 23-18 is made in Paskievici [6]. As a result, the best value for the number of occupational

Table 23-17
Occupational Injuries from Accidents, Natural-Gas Cycle

	Step				
Author	Extraction	Processing	Transport	Conversion	Total
WASH-1224	3.33	0.75	1.73	2.0	7.81
Hamilton	16.0	0.07	1.6	1.47	19.14
Hittman	1.1	0.16	1.23	4.5	7.0
Smith et al.[a]	40-110	1.4-2.7	13-48	3.9	58-165
Comar and Sagan	3.33-28.0	0.07-0.75	1.6-1.73	0.8-2.0	5.80-32.5
MITRE	21.28	0.07	1.6	1.45	24.4
Inhaber (1978)	4-31	←	1.7-1.9	0.9-2.1	6-35
AMA					5.3-32.0
Inhaber (1980)	4-31	←	1.7-1.9	0.9-2.1	13-41[a]
CONAES					24.4

[a]The total includes 6 injuries that occur during construction.

Table 23-18
Occupational Fatalities from Accidents, Uranium Cycle

	Step				
Author	Extraction	Processing	Transport	Conversion	Total[a]
WASH-1224	0.12	0.007	0.003	0.013	0.13
Hamilton	0.27	0.07	—	0.013	0.35
Hittman	0.16	←	←	0.015	0.175
Smith et al.	0.12-0.27	0.007	0.003-0.012	0.013-0.017	0.14-0.31
Comar and Sagan	0.07-0.27	0.004-0.27	0.003	0.013	0.09-0.55
OECD	0.3	0.12	0.003	0.27[b]	0.69[b]
Morris (1977)	0.12-0.27	0.005	—	0.013	0.13-0.28
WHO	0.1	0.13	< 0.01	0.27[b]	0.55[b]
Inhaber (1978)	0.08-0.57	←	0.003-0.012	0.013-0.017	0.10-0.60
AMA	0.07-0.27	0.004-0.27	0.003-0.007	0.013	0.08-0.55
United Kingdom	0.1	—	—	0.15	0.25
Ramsay					0.09-0.31[b]
Schurr et al.					0.08-0.27[b]
Inhaber (1980)	0.12-0.57		0.003-0.012	0.013-0.017	0.25-0.75[b]
CEPN	0.043	0.129	0.009	0.018	0.19
CONAES	0.26	0.001	0.013	0.013	0.27

[a]The total is not always equal to the sum because of roundings.
[b]The total includes accidents that occur during construction.

fatalities has been taken as 0.20(2). As mentioned before, the factor of uncertainty refers mainly to the differences in the approaches taken by the authors.

Occupational Injuries. Table 23-19 shows the estimates of occupational injuries. The previous remarks apply also for this category of accidents. The best value was taken as 15(2).

Risk Incurred by the General Population. The only risks for the general population can be related to transport of uranium, of spent fuel, and of radioactive wastes. These risks are considered small or negligible.

Summary and Discussion

The comparative study of accident hazards arising from different fuel cycles indicates that reasonable agreement, within a factor of 2, exists between estimates made by most of the authors, and that the uncertainty factor is mainly due to different working conditions and to interpretation of statistical data. Although it is recognized that the omission of risks associated with hypothetical risks is debatable, our best estimates of real

Table 23-19
Occupational Injuries from Accidents, Uranium Cycle

Author	Extraction	Preparation	Transport	Conversion	Total[a]
WASH-1224	4.8	2.05	0.06	1.73	8.65
Hamilton	13.3	0.8	—	1.73	15.83
Hittman	4.63	←	—	1.49	6.12
Smith et al.	3.2-13.3	2.0	0.06-0.12	1.7	7.1-16.9[b]
Comar and Sagan	2.4-13.3	0.8-2.0	0.06-0.19	1.73	5.0-17.2
Inhaber (1978)	3.4-16.0	←	0.06-0.20	1.7	5.0-18.0[b]
AMA	2.4-13.3	0.8-2.0	0.06-0.19	1.73	5.0-17.2
Ramsay					16-32[b]
Schurr et al.					13-26[c]
Inhaber (1980)	3.4-16.0	←	0.06-0.20	1.7	16-29[b,c]
CEPN	29.2	7.9	2.1	6.5	45.7
CONAES					20.0[c] (?)

[a]The total is not always equal to the sum because of roundings.
[b]The total includes 0.12 injuries that occur during the waste disposal.
[c]The total includes injuries that occur during construction: 9-18 (Ramsay and Schurr et al.) or 11 (Inhaber 1980).

risks from each fuel cycle and for each category have been put together in table 23-20. From this table the following general conclusions may be drawn:

1. The risk of occupational fatalities is very small and similar for the natural-gas and for the uranium cycles; for the oil and coal cycles this risk is about twice and seven times larger, respectively. For the coal cycle the major risk occurs during the extraction step, whereas for the uranium cycle the risk is about equally divided among extraction and conversion steps if accidents during the construction phase of the nuclear power plant are included, as is done by some authors.
2. The risk of fatalities in the general population is in general negligible for the oil, natural-gas, and uranium cycles; but it is about 100 times higher for coal than for natural gas and uranium. In all cases the risk is related to the transport of the fuel.
3. The risk of occupational injuries shows the same characteristics as the risk of occupational fatalities: it is lowest for natural gas and uranium, about twice as high for oil, and four times higher for coal.
4. The risk of injuries in the general population is very low; for coal, it is of an order of magnitude higher than for gas and uranium.

**Table 23-20
Comparison of Accident Hazards Associated with Different Fuel Cycles Normalized to a Production of GW-Year of Electrical Energy**

Fuel Cycle	Workers		Population		Total	
	Fatal	Nonfatal	Fatal	Nonfatal	Fatal	Nonfatal
Coal	1.40(1.5)	60(1.5)	1.0(1.5)	1.8(2.0)	2.40(1.5)	62(1.5)
Oil	0.35(1.5)	30(1.5)	?	?	0.35(1.5)	30(1.5)
Gas	0.20(1.5)	15(2)	0.009	0.005	0.21(1.5)	15(2)
Uranium	0.20(1.5)	15(2)	0.012	0.11	0.21(1.5)	15(2)

Note: Best estimates are accompanied by uncertainty factors, in parentheses.

References

1. Hull, A.P. "Radiation in Perspective: Some Comparison of the Environmental Risks from Nuclear and Fossil Fueled Power Plants." *Nuclear Safety* 2, no. 3 (May-June 1971).
2. U.S. Atomic Energy Commission. "Comparative Risk-Cost -Benefit Study of Alternative Sources of Electrical Energy: A Compilation of Normalized Cost and Impact Data for Current Types of Power Plants and their Supporting Fuel Cycles." WASH-1224 (1974).
3. Hamilton, L.D. "Risk and Consequences in Energy Production." Proceedings of the First International Conference on Health Effects of Energy Production, Chalk River Nuclear Laboratories, Ontario (12-14 September 1979). AECL-6958, Atomic Energy of Canada, Ltd.
4. Morris, G. "Problems and Pitfalls in the Environmental Assessment of Energy Systems." Proceedings of the First International Conference on Health Effects of Energy Production, Chalk River Nuclear Laboratories, Ontario (12-14 September 1979). AECL-6958, Atomic Energy of Canada, Ltd.
5. Paskievici, W. "Evaluation des risques associés aux chaînes énergétiques—Aspects méthodologiques." Proceedings of the 1980 Annual Conference of the Canadian Nuclear Society, held in Montreal, Quebec, Canada, June 18th, 1980.
6. Paskievici, W. "La sécurité des modes de production d'électricité." Unpublished report for Hydro-Quebec. Available (in draft form) from Institut de Génie Nucléaire, Ecole Ploytechnique, Montreal, Quebec.
7. Inhaber, H. "Risk of Energy Production." Atomic Energy Control Board, Report AECB-1149, revised (1978).
8. Hamilton, L.D. "The Health and Environmental Effects of Electricity Generation—A Preliminary Report." Report BNL-20582, Brookhaven National Laboratory, Upton, N.Y. (1974).

9. Hittman Associates, Inc. "Environmental Impacts, Efficiency and Cost of Energy Supply and End Use." Report prepared for the National Science Foundation, the Environmental Protection Agency, and the Council on Environmental Quality. NTIS Report PB-239 159 (1974).
10. Smith, K.R., et al. "Evaluation of Conventional Power Systems." Report ERG 75-7, University of California, Berkeley (1975).
11. Comar, C.L., and Sagan, L.A. "Health Effects of Energy Production and Conversion." *Annual Review of Energy* 1 (1976):581-600.
12. Pochin, E.E. "Estimated Population Exposure from Nuclear Power Production and Other Radiation Sources." Nuclear Energy Agency, OCDE, Paris (1976).
13. Mitre Corporation. "Accidents and Unscheduled Events Associated with Non-nuclear Energy Resources and Technology." Report prepared for the Environmental Protection Agency. NTIS Report PB-265 398 (1977).
14. Morris, S.C. "Comparative Effects of Coal and Nuclear Fuel on Mortality." Report BNL-23 579, Brookhaven National Laboratory (1977).
15. World Health Organization. "Health Implications of Nuclear Power Production." Report of a Working Group, WHO, Regional Office for Europe, Copenhagen (1978).
16. Institute for Energy Analysis, ORAD. *Economic and Environmental Impacts of a U.S. Nuclear Moratorium, 1985-2000.* Cambridge, Mass.: MIT Press, 1979.
17. AMA Council on Scientific Affairs. "Health Evaluation of Energy-Generating Sources." *J. Amer. Med. Assoc.* 240 (1978):2193-2195.
18. "The Hazards of Conventional Sources of Energy." Report by the Health and Safety Commission. London: Her Majesty's Stationery Office, 1978.
19. Ramsay, W. "Unpaid Costs of Electrical Energy: Health and Environmental Impacts from Coal and Nuclear Power." Baltimore, Md.: Johns Hopkins University Press, 1978.
20. Schurr, S.H., et al. "Energy in America's Future: The Choices Before Us." A study prepared for the RFF National Energy Strategies Project, published for Resources for the Future. Baltimore and London: Johns Hopkins University Press, 1979.
21. Morris, S.C.; Novak, K.M.; and Hamilton, L.D. "Databook for the Quantitation of Health Effects from Coal Energy Systems." Draft, Brookhaven National Laboratory, Upton, N.Y. (1979).
22. Inhaber, H. "Risk of Energy Production." Report AECB/REV-3, Atomic Energy Control Board, Ottawa (1980).
23. Belhoste, J.F.; Durant, B.; and Maccia, C. "Risques sanitaires et écologiques de la production d'énergie électrique. Cycle nucléaire

(PWR), fuel, charbon." Report 20-3. Présentation des scénarios, Centre d'etudes sur l'évaluation de la protection dans le domaine nucléaire, Fontenay-aux-Roses, France (1979).
24. CONAES (Committee on Nuclear and Alternative Energy Systems). "Alternative Energy Demand Futures." U.S. National Academy of Sciences (1980).
25. National Academy of Sciences, National Academy of Engineering, and National Research Council. "Air Quality and Stationary Source Emission Control." Report prepared for the Committee on Public Works, U.S. Senate (March 1975).
26. Holdren, J.P., et al. "Risk of Renewable Energy Sources: A Critique of the Inhaber Report." Report ERG 79-3, Energy and Resources Group, University of California, Berkeley (1979).
27. Morgan, M.G., et al. "A Probabilistic Methodology for Estimating Air Pollution Health from Coal-Fired Power Plants." *Energy Systems and Policy* 2, no. 3 (1978):287-310.
28. Buehring, W.A.; Foell, W.K.; and Keeney, R.L. "Energy/Environment Management: Application of Decision Analysis." Research Report RR-76-14, International Institute for Applied System Analysis (IIASA), Laxenburg, Austria (1976). Cited in [19].
29. U.S. Nuclear Regulatory Commission. "Reactor Safety Study: An Assessment of Accident Risks in U.S. Commercial Nuclear Power Plants." WASH-1400 (1975).
30. "Risk Assessment of Storage and Transport of Liquefied Natural Gas and LP-Gas." Science Applications, Inc. (1974). Reported in [13].
31. U.K. Health and Safety Executive. "Canvey: An Investigation of Potential Hazards from Operations in the Canvey Island/Thurrock Area." London: Her Majesty's Stationery Office, 1978.

24 Environmental Risks of Energy Production: The Carbon Dioxide Example

Robert S. Chen

Introduction

All energy systems pose risks, and in recent years we have begun to discover how risky indeed some of them may be. We are also learning that there are many different kinds of risks, as well as benefits, that make choices among energy systems extremely difficult. As energy demand increases in magnitude beyond levels that can be met with conventional energy sources, careful consideration of the risks and benefits of energy alternatives, and the distribution and unique characteristics of these risks and benefits, is becoming increasingly important.

One risk that has received considerable attention in recent years is the possibility that the combustion of fossil fuels may substantially increase carbon dioxide (CO_2) concentrations in the atmosphere, which might in turn alter the global climate. Such alteration could have important impacts, both beneficial and adverse, on human activities and welfare, especially for food production, water-resource management, and energy supply and demand. This risk, though uncertain in likelihood and potential consequence, is nevertheless important, as it represents a credible global threat to the environment in which humanity has lived for millennia.

This chapter briefly reviews the present understanding of the so-called carbon dioxide problem, including the origin of the problem, its climatic and environmental aspects, and its potential implications for society. I will draw freely from a number of studies conducted by my organization, the National Academy of Sciences. However, any views and interpretations are strictly my own. For more detailed reviews, see Schneider and Chen (1980) or Marland and Rotty (1979).

Carbon Dioxide in the Atmosphere

Concern over the rise of CO_2 in the atmosphere dates back at least to Arrhenius (1896), but only in recent years has the possibility gained wide-

---

The views expressed in this paper are those of the author and do not necessarily represent the views of the Climate Board or the National Academy of Sciences.

spread attention. Much of this attention arises because of a twenty-year record of careful measurements of CO_2 concentrations in the atmosphere at Mauna Loa, Hawaii, made by Keeling et al. (1976b) (figure 24-1). This curve is one of the few unequivocal facts in the CO_2 problem: the fact is that atmospheric CO_2 has been increasing steadily for at least the past two decades at about 1-1.5 parts per million (ppm) by volume per year, to the present level of about 335 ppm. Superimposed on this trend are a seasonal cycle of about 5 ppm and an interannual variation of about 1 ppm. The seasonal cycle is generally believed to result from the expansion and contraction of the biota in the Northern Hemisphere (Bolin and Keeling 1963); the smaller interannual variation may be associated with a climatic phenomenon known as the southern oscillation (Bacastow 1976, 1977; Rust, Rotty, and Marland 1979). Similar trends are found in measurements at the South Pole, at high altitudes, and even in highly polluted urban atmospheres (Keeling et al. 1976a; Bolin and Bischof 1970; MacRae and Graedel 1979). Moreover, on the basis of actual measurements and proxy data, it is believed that concentrations have been rising steadily since the mid- to late nineteenth century, when CO_2 levels may have been as much as 45-70 ppm less than present levels (Callendar 1958; Stuiver 1978a,b; Keeling 1978).

Where does this CO_2 come from? As figure 24-2 shows, human emissions of CO_2 primarily the combustion of fossil fuels, have also increased steadily since the mid-nineteenth century except for two short breaks during World War I and World War II (Rotty 1979). Before jumping to conclusions, however, note that if all the emissions shown here remained in the atmosphere, measured levels would be increasing twice as fast as they are (for example, see Keeling and Bacastow 1977). That is, at most only about 50 percent of the CO_2 released from fossil fuels by humans stays in the atmosphere. The problem is thus quite complex—as indicated earlier, atmospheric CO_2 levels are influenced strongly by fluctuations in the biota and other parts of the climate system, such as the oceans. In fact, no one is really sure where the other 50 percent or so of the emitted CO_2 goes, although recently it has been suggested that much more of it may reach the deep layers of the oceans than previously believed (Thompson and Schneider 1979; Climate Research Board 1979). Considerable uncertainty also exists in our knowledge of the biota's role—we do not even know if the biota are a net source or sink of CO_2. Some biologists, for example, believe that extensive tropical deforestation, land clearing, and possibly soil erosion may be adding substantial amounts of CO_2 to the atmosphere (Woodwell et al. 1978; Stuiver 1978a; Bolin 1977; Adams, Mantovani, and Lundell 1977). Thus the CO_2 problem may be more than just a question of fossil-fuel energy consumption; it may also involve agricultural and land-use practices.

Let me backtrack one step. As indicated by the logarithmic scale in

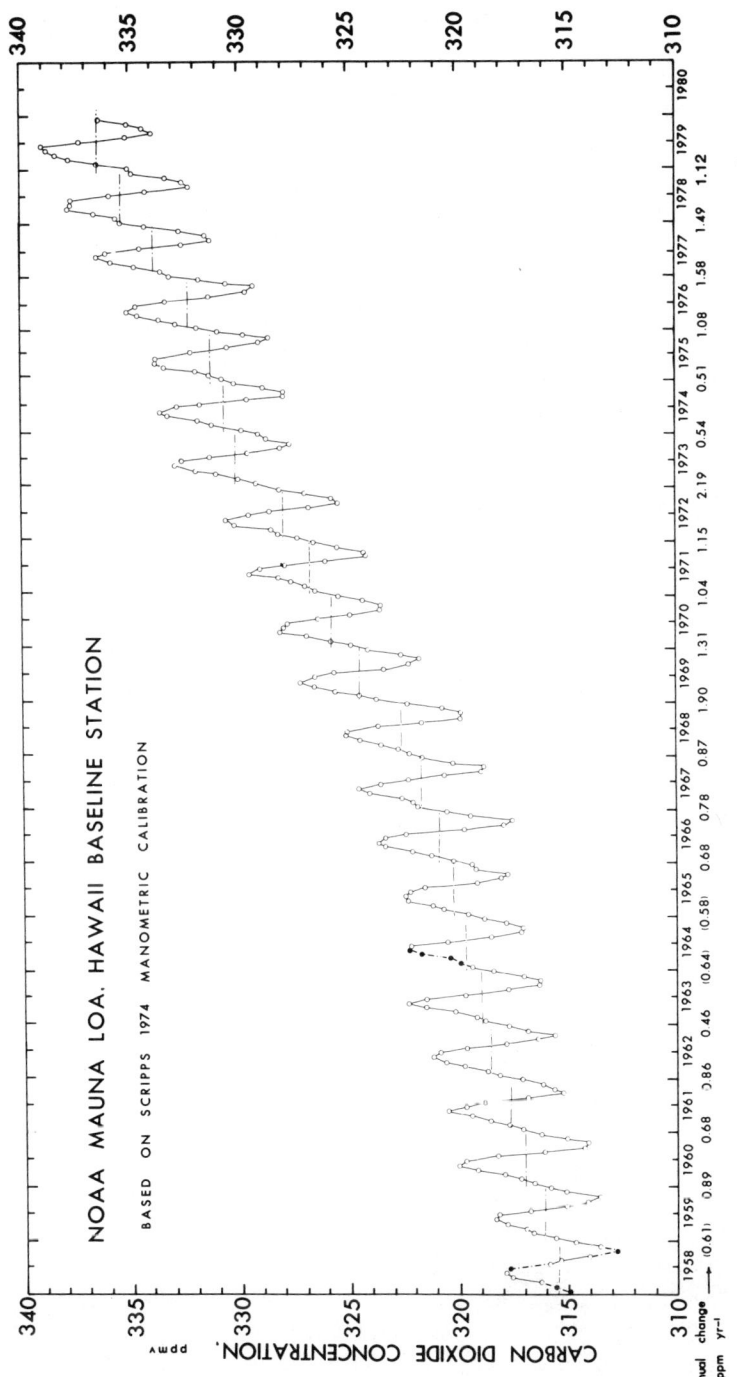

Figure 24-1. Concentration of Atmospheric CO_2 Measured at Mauna Loa Observatory, Hawaii (10.5°N, 155.6°W)

Source: Courtesy of W. Elliott, Air Resources Laboratory, National Oceanic and Atmospheric Administration.
Note: The circles represent the observed monthly average concentration based on continuous measurements; the black dots indicate months in which some data are questionable.

figure 24-2, there is a steady increase in the rate of CO_2 emissions, presumably arising from the combination of rapid population growth and increasing per capita energy use. Until the oil embargoes of the early 1970s, the annual growth in energy consumption was over 4 percent; this has lessened somewhat in recent years. As long as growth in fossil fuel energy use remains exponential, however, it seems quite likely that atmospheric CO_2 concentrations will increase substantially—by at least 50 percent, and probably 100 percent, within 50-100 years (for example, see MacDonald 1978). The 50-year estimate means that the half or so of the world's population that is now under 25 years of age will have a good chance of witnessing these higher levels—this is not just a problem for our descendants.

An important question is, of course, whether the fossil fuels are available—and available at tolerable economic and environmental costs—to sustain exponential growth. It turns out there is at least enough carbon buried primarily in known coal reserves to raise the atmospheric content of CO_2 some six to seven times above the preindustrial level (about 300 ppm). Interestingly, as table 24-1 indicates, about 90 percent of the estimated remaining carbon is located in the United States, the USSR, and China (Ausubel 1980; see also Perry and Landsberg 1977; Workshop on Alternative Energy Strategies 1977). It is also worth noting that different fossil fuels—natural gas, oil, coal, and now synthetics—result in different amounts of CO_2 released for the same amount of energy output (for example, see MacDonald 1978; Sundquist and Miller 1980). The difference can be as much as a factor of two or three if the extra energy to convert and deliver fuels in usable form is taken into account. However, although it may make some sense to use less CO_2-intensive fuels, such as natural gas, as much as possible, net CO_2 levels are still likely to increase as long as population and per capita energy use continue to increase.

Environmental Impacts of Increasing Carbon Dioxide

What if CO_2 concentrations increase by half or more over present levels? There may be some direct beneficial effects: plants will have more CO_2 to photosynthesize and may therefore be able to grow faster. Of course, even though enhancement of CO_2 levels is a technique often used in greenhouses to accelerate plant growth, there are likely to be other limitations in water, nutrients, and sunlight that constrain actual productivity gains in the field. This direct effect of CO_2 is being studied closely (Carbon Dioxide Effects Research and Assessment Program 1980).

Another possible effect of increased CO_2 levels has recently come to light: the possibility of health effects attributable to prolonged exposure to high CO_2. Very limited animal experiments have shown some decalcifica-

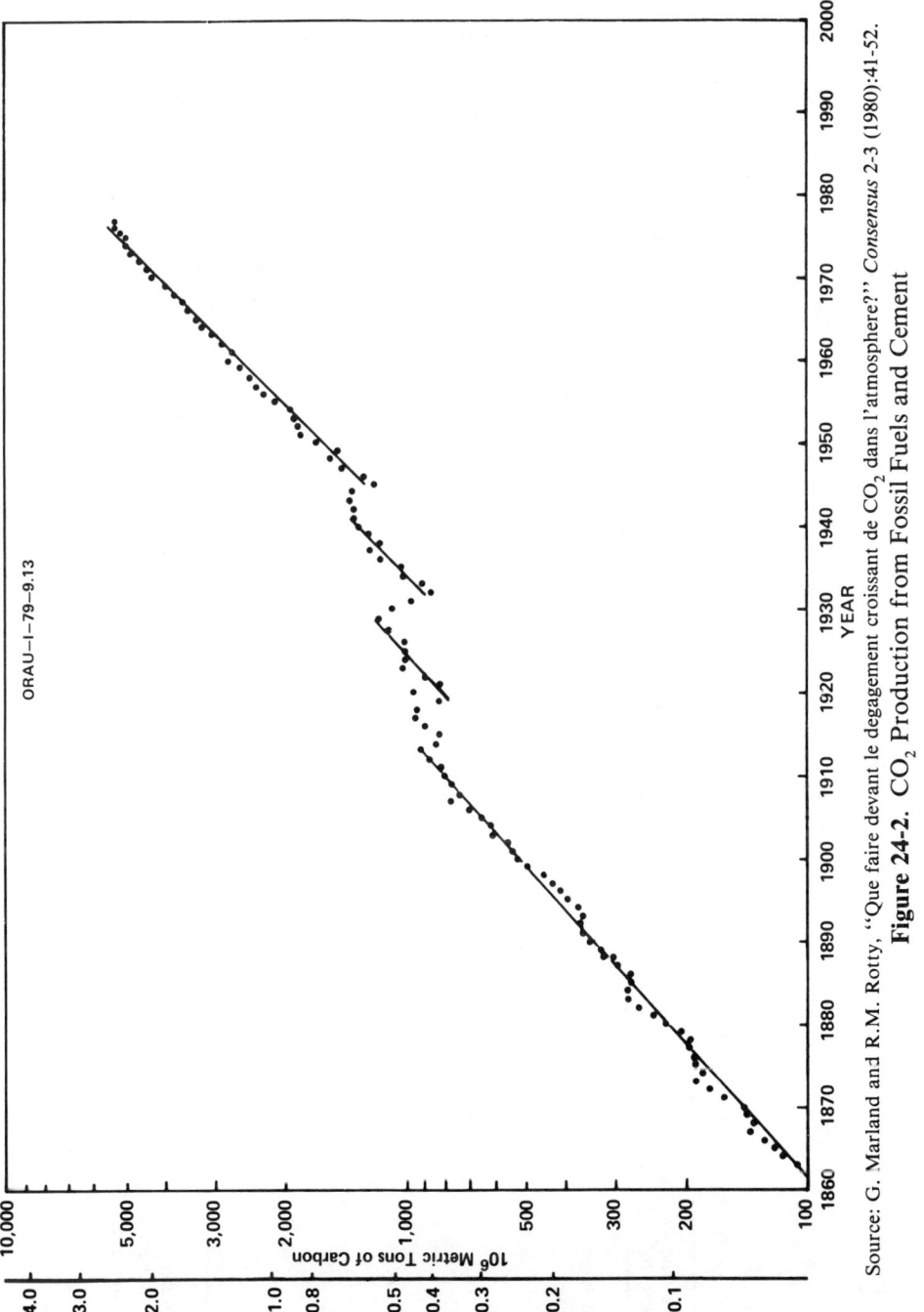

Source: G. Marland and R.M. Rotty, "Que faire devant le degagement croissant de CO$_2$ dans l'atmosphere?" *Consensus* 2-3 (1980):41-52.

Figure 24-2. CO$_2$ Production from Fossil Fuels and Cement

Table 24-1
World Distribution of Coal Resources

Greater than 10^{12} tce ($1,000 \times 10^9$ tce)		Between 10^{11} and 10^{12} tce (100 and $1,000 \times 10^9$ tce)		Between 10^{10} and 10^{11} tce (10 and 100×10^9 tce)		Between 10^9 and 10^{10} tce (1 and 10×10^9 tce)	
USSR	4,860	Australia	262	India	57.0	East Germany	9.4
United States	2,570	West Germany	247	South Africa	57.0	Japan	8.5
China	1,438	United Kingdom	163	Czechoslovakia	17.5	Colombia	8.3
		Poland	126	Yugoslavia[a]	10.9	Zimbabwe	7.1
		Canada	115	Brazil	10.0	Mexico	5.5
		Botswana	100			Swaziland	5.0
						Chile	4.6
						Indonesia[a]	3.7
						Hungary[a]	3.5
						Turkey	3.3
						Netherlands	2.9
						France	2.3
						Spain	2.3
						North Korea	2.0
						Romania	1.8
						Bangladesh	1.6
						Venezuela	1.6
						Peru	1.0

Source: J. Ansubel, "Economics in the Air—An Introduction to Economic Issues of the Atmosphere and Climate," in *Climatic Constraints and Human Activities*, J. Ausubel and A.K. Biswas eds. (New York, Pergamon, 1980), pp. 13-59, based on data after the World Energy Conference, 1978. Reprinted with permission.

Note: Units are 10^9 tons of coal equivalent (tce).

[a]Mostly lignite.

tion of bones and accumulation of minerals in the kidney at levels of about 1,000 ppm. Submariners exposed to much higher levels than this, but for only a few months, have had similar problems. These effects have not yet been closely examined (L. Machta, Air Resources Laboratory, National Oceanic and Atmospheric Administration, personal communication, 1980).

What has been studied in considerable detail is the potential effect of increased CO_2 levels on the earth's climate. By *climate* I mean the atmospheric conditions—the weather—over some period of time, usually months or longer. As early as a century ago, scientists recognized the importance of CO_2 in the radiation balance of the earth's atmosphere (see figure 24-3). That is, CO_2 absorbs infrared radiation emitted by the earth's surface that would otherwise be lost to space. Increased CO_2 thus means that more heat is trapped in the atmosphere, raising its temperature on average. This effect is often somewhat misnamed the greenhouse effect. An important feedback is that the added heat absorbed by CO_2 causes slightly more evaporation of water; the added moisture in the atmosphere in turn absorbs more radiation, augmenting the CO_2 effect. A 100-percent increase in CO_2 may thus result in a global average surface warming of some 3-4°C, including the moisture feedback.

This simple mechanism is the basis of our belief that higher CO_2 levels could have substantial climatic effects. Things are not quite this simple, of course, as there are other feedback mechanisms that could reduce or increase a CO_2-induced warming. For example, more moisture implies more clouds; but depending on the type of cloud, the change in the albedo or reflectivity of the earth could be either positive or negative (for example, see Schneider, Washington, and Chervin 1978). However, to the best of our knowledge, all known, important feedback mechanisms have been incorporated into three-dimensional computer models of the general circulation of the atmosphere; these models support the basic mechanism I have described (Climate Research Board 1979). These models also suggest that a CO_2-induced climate warming would be accompanied by major changes in precipitation patterns, winds, and other climatic parameters. Moreover, depending on the latitude, season, or other factors, actual local increases in temperature could be more or less than the global average increase—perhaps, for example, three to five times larger in polar regions (Manabe and Wetherald 1975; Schneider 1975). Unfortunately, our models are not yet able to make reliable, time-dependent, regional predictions of climatic changes. One important failing is that they lack a good representation of the world's oceans; as suggested by one academy report (Climate Research Board 1979), the oceans may be a "great and ponderous flywheel" that will absorb CO_2 and heat, thereby delaying—but not counteracting—a temperature increase (see also Thompson and Schneider 1979). Therefore, we may not begin to experience a climate warming until well after it is too late to do anything about it.

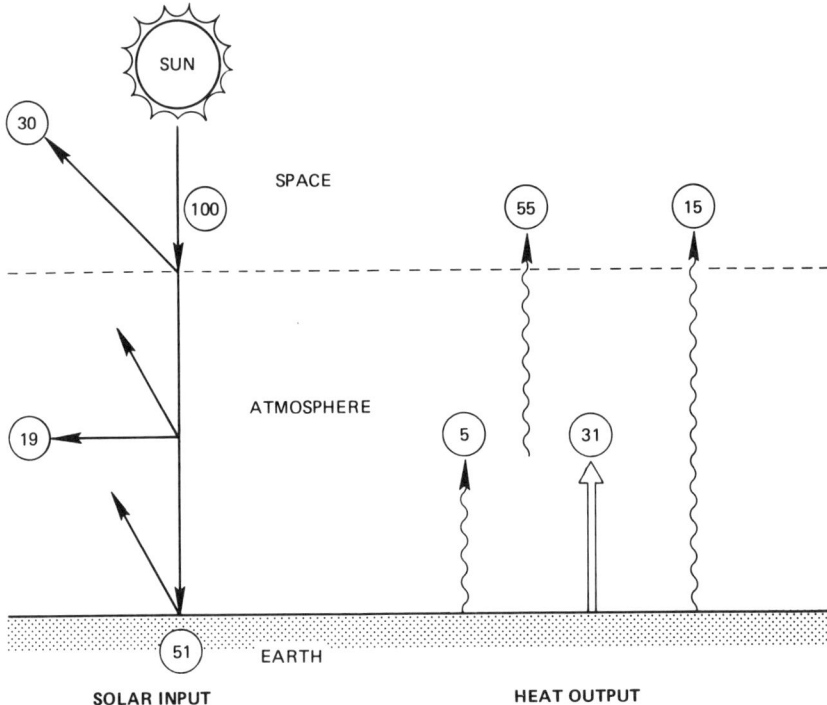

Source: Courtesy of J. Peny, after G.W. Paltridge and C.M.R. Platt, *Radiative Processes in Meteorology and Climatology*, Developments in Atmospehric Science, vol. 5 (New York: Elsevier, 1976), figure 1.4.

Note: Solid arrows represent radiant solar energy; wavy arrows show heat (infrared) radiation; the double arrow represents heat transferred to the atmosphere by convection and evaporation. Numbers indicate percentages of mean solar input (100 = 34 mW/cm^2). Note that the heat output to space (55 + 15) plus the reflected radiation (30) equals the solar input, the heat output of the atmosphere (55) equals the amount absorbed (19 + 5 + 31), and the heat output of the earth (5 + 31 + 15) equals the amount absorbed (51).

Figure 24-3. Simplified Schematic Diagram of the Earth's Energy Balance in the Global and Multiannual Mean

What is the significance of a global average warming of 3°C? By way of comparison, the climate is believed to have been several degrees Centigrade warmer some 120,000 years ago. This may have corresponded to a sea level about 6 meters higher than at present, for reasons that will be explained shortly (for example, see Mercer 1968, 1978). More recently, an average temperature less than 1°C warmer than now in the Northern Hemisphere may have been an important factor in the dust bowl era of the 1930s. In the opposite direction, the so-called little Ice Age of the fifteenth, sixteenth, and seventeenth centuries was only about 1.5°C cooler on average than now. Extensive glaciation corresponds to temperatures some 10°C cooler. Thus a temperature change of several degrees Centigrade is significant even on glaciological time scales (for example, see Schneider with Mesirow 1977; Mitchell 1977).

Societal Implications of Climatic Change

It is evident that a climate warming could have both beneficial and adverse impacts on society. A warmer climate could mean longer growing seasons, lower heating needs, and increased precipitation on average. Conversely, it could also mean greater prevalence of certain plant, animal, and human pests and diseases; increased air-conditioning demands; and altered precipitation and evaporation patterns (for example, see Kellogg and Schware 1980). Indeed, some changes are likely to be viewed as beneficial by some and adverse by others. For example, higher summer temperatures may lead to better business for refrigeration and air-conditioning equipment manufacturers, but higher cooling costs for consumers. It is therefore important to consider the distribution of costs and benefits for different regions, groups, and even time periods. Overall, however, climatic changes could severely stress the global food system, which at present can barely keep pace with growing population and per capita demand. Such changes could also worsen the degradation of the environment, for example, by accelerating desertification or further straining sensitive ecosystems. They could significantly alter energy-demand patterns and, in the case of renewable energy resources, alter the availability of energy. Many of these impacts on society may be manifested as increased poverty, hunger, sanitation and health problems, migration, international tension, and so forth (Chen 1980; Panel IV 1980; Climate Research Board 1980).

One problem that we face is that these climate impacts are hard to quantify. Without precise predictions of where, when, and how climate may change, it is difficult to assess what the direct effects on the environment and society might be, in terms of, say, crop-yield gain or loss, altered energy demand, or modified water supply. Indeed, even for present-day climate fluctuations—that is, the normal variation in the weather from day to day, month to month, or even year to year—we do not have good estimates of costs and benefits. We do know they can be large, however. For example, the recent heat-wave/drought combination in the United States inflicted on the order of $15-20 billion in losses, according to preliminary estimates (Center for Environmental Assessment Services 1980). Notably, despite major grain losses among most farmers, winter-wheat farmers benefited from the drought and are likely to have a bumper crop this year.

Assessments of these kinds of impacts are made even more difficult because of interactions between environmental and societal factors. For example, unexpected and extensive damage was reported in the Midwest as a result of the combination of extreme drought followed by extreme wetness—this caused many in-ground structures such as sewers and storm drains to buckle or crack (N. Rosenberg, Univ. of Nebraska, personal communication, 1980). Another example is that some people in high-crime areas may have suffered even more from the heat because they were apparently afraid to open their windows—a synergism between a climatic extreme and the societal problem of crime (Hock 1980).

There is one possible consequence of a climate warming in which the impacts can be quantified to some degree at present. This is the possibility that a CO_2-induced warming, amplified, as mentioned earlier, several times in polar regions, could initiate the collapse of the western portion of the ice cap in the Antarctic (see figure 24-4), causing a worldwide rise in sea level of some 5-6 meters. Some glaciologists believe that this so-called West Antarctic Ice Sheet may be extremely unstable, so that a warming of 5-10°C could trigger its rapid collapse—where "rapid" to a glaciologist is probably on the order of decades to many centuries (Mercer 1978; Hughes 1973; Hollin 1969, 1972; Thomas, Sanderson, and Rose 1979). Though speculative, this possibility is at least plausible in light of our present limited knowledge. Moreover, as also mentioned earlier, there is extensive evidence that the sea level was some 6 meters higher than at present some 120,000 years ago during a warm interglacial period (Mercer 1968; Neumann and Moore 1975; Aharon, Chappell, and Compston 1980). Indeed, some scientists have suggested that the deglaciation of the East Antarctic Ice Sheet is possible, leading to a sea-level rise ten times greater than discussed here (Wilson 1969; Hollin 1969, 1972, 1980).

What then would be the impacts of a 5-6 meter rise in sea level? One of my colleagues, Dr. Stephen Schneider, and I looked closely at the permanent coastal flooding that would result in the United States if indeed a rise were to occur (Chen and Schneider, in press; see also Schneider and Chen 1980). Figure 24-5 is a map of Florida. The light grey areas are the land below the 15- and 25-foot contours of elevation above present sea level. About one-third of Florida would be inundated by a 25-foot rise, including most of the major cities. Louisiana would be in about the same boat, as figure 24-6 shows. The mid-Atlantic, north Atlantic, and West Coast regions (figures 24-7 through 24-9) would be affected to a lesser, though still significant, extent. Even mild hurricanes like Hurricane Jeanne can cause extensive flooding and substantial damage (for example, see White and Haas 1975). Of course, the kind of coastal flooding I am talking about would probably occur over decades at a minimum, so it is likely that everyone could safely move away. Nevertheless, our calculations show that order-of-magnitude

Environmental Risks of Energy Production 285

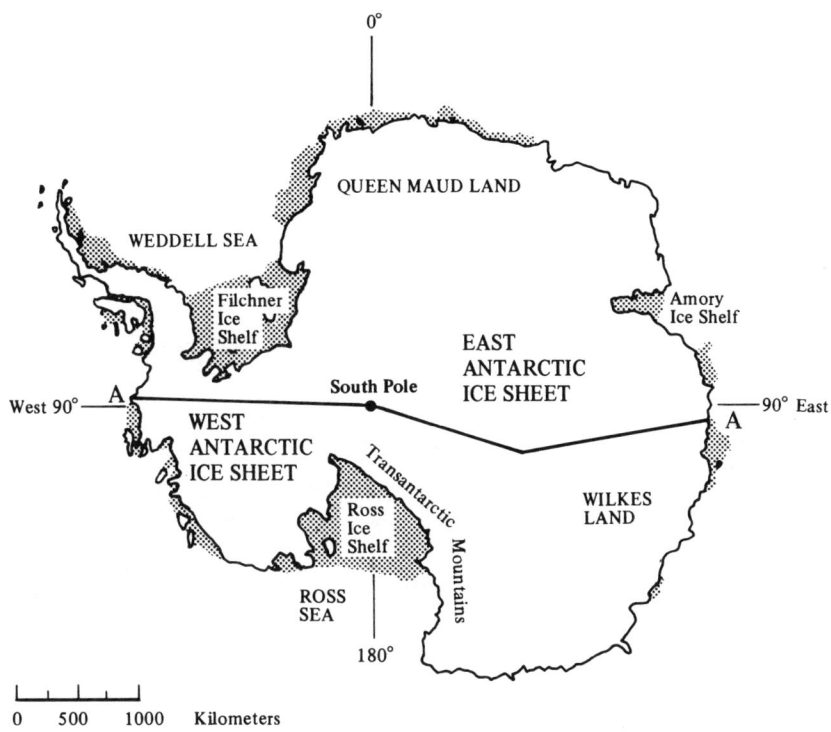

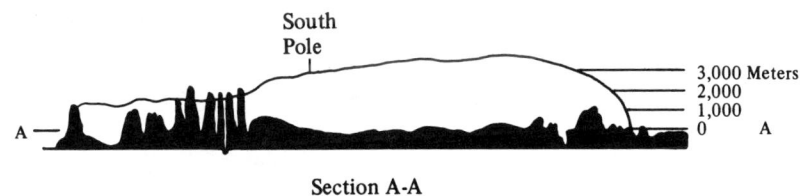

Section A-A

Source: After G.H. Denton, R.L. Armstrong, and M. Stuiver, "The Late Cenozoic Glacial History of Antarctica," in *The Late Cenozoic Glacial Ages*, K. Turekian, ed. (New Haven: Yale University Press, 1971), pp. 267-306.

Figure 24-4. Map of the Antarctic Ice Cap and a Cross-Section

economic losses could be substantial—in the case of Florida, some 40 percent of its population and half its total value in land and structures would be affected by a 15-foot rise, based on 1970 and 1971 data (Bureau of the Census 1973a,b). Nationwide, on the order of $220 billion (1971 dollars) in land and structures could be lost to a 25-foot rise (see table 24-2). The sites of at

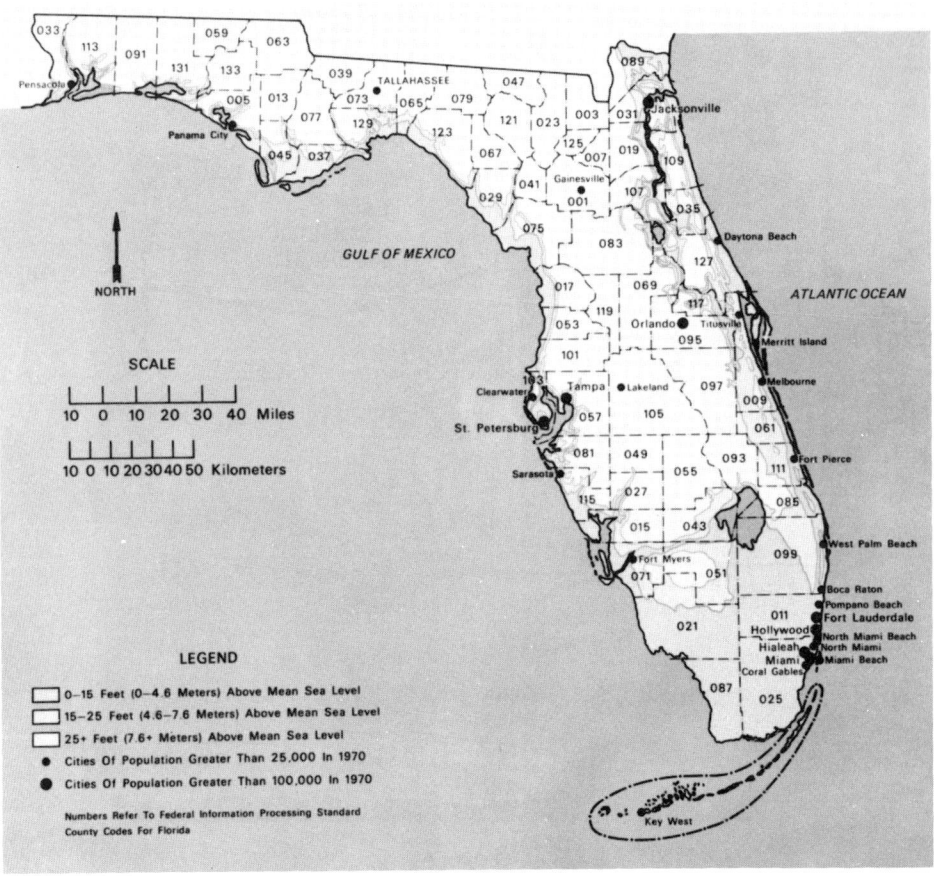

Source: S.H. Schneider and R.S. Chen, "Carbon Dioxide Warming and Coastline Flooding: Physical Factors and Climatic Impact," *Ann. Rev. Energy* 5 (1980):107-140. Reprinted with permission.

Figure 24-5. Approximate Areas Flooded by a 15-25-Foot (4.6-7.6-Meter) Rise in Mean Sea Level, Florida

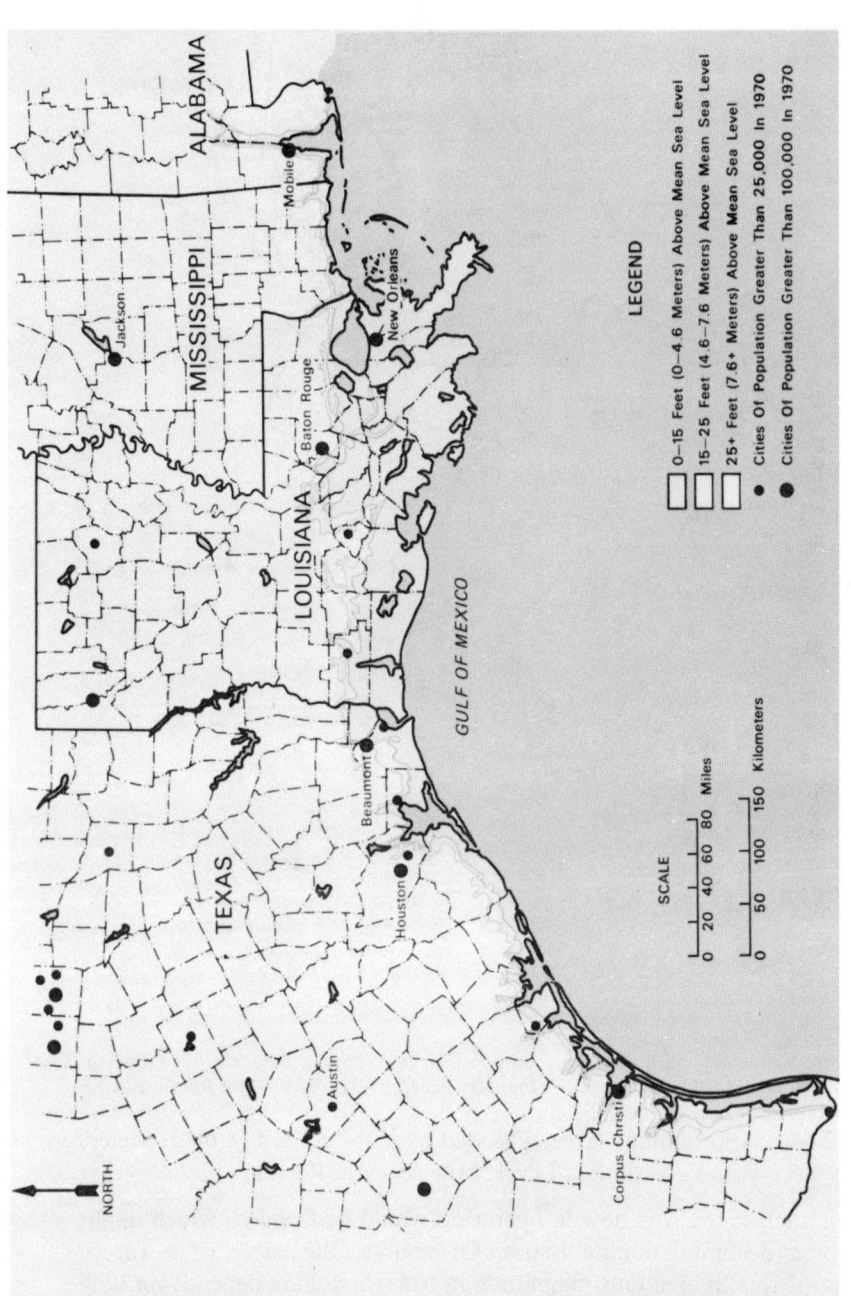

Source: S.H. Schneider and R.S. Chen, "Carbon Dioxide Warming and Coastline Flooding: Physical Factors and Climatic Impact," *Ann. Rev. Energy* 5 (1980):107-140. Reprinted with permission.

Figure 24-6. Approximate Areas Flooded by a 15-25 Foot (4.6-7.6-Meter) Rise in Mean Sea Level, Gulf Coast Region

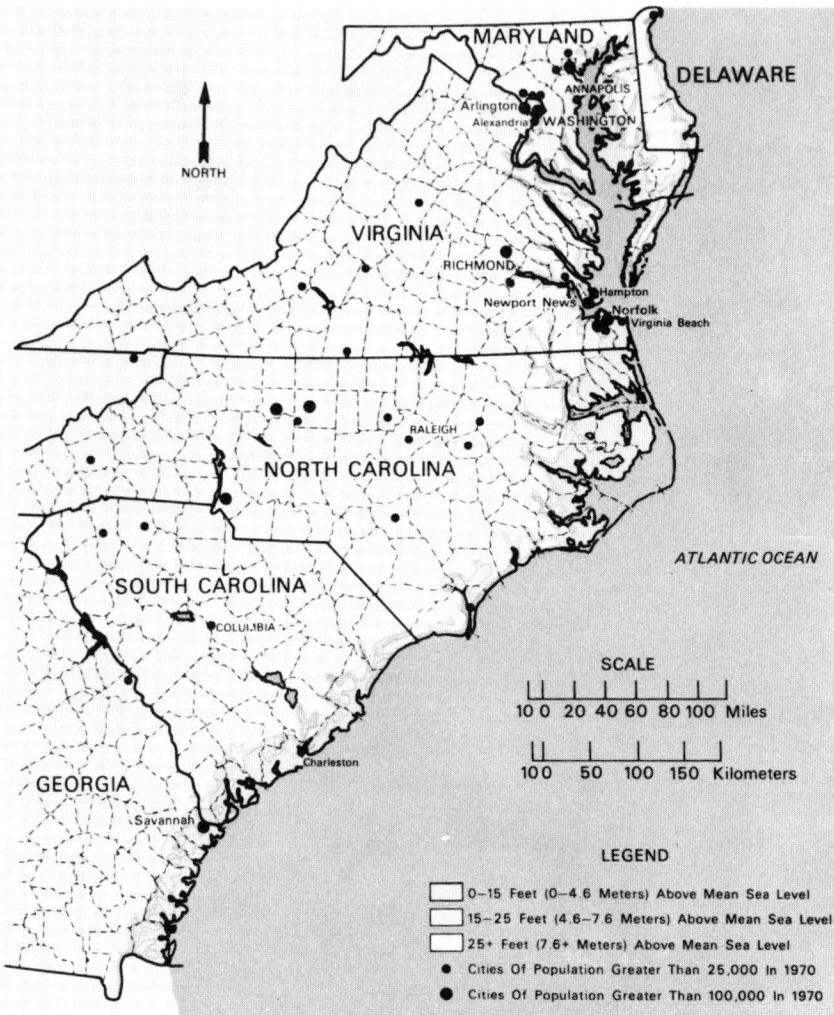

Source: S.H. Schneider and R.S. Chen, "Carbon Dioxide Warming and Coastline Flooding: Physical Factors and Climatic Impact," *Ann. Rev. Energy* 5 (1980):107-140. Reprinted with permission.

Figure 24-7. Approximate Areas Flooded by a 15-25-Foot (4.6-7.6-Meter) Rise in Mean Sea Level, Mid-Atlantic Region

least ten nuclear reactors now in operation would be flooded, which might raise some decommissioning issues. Of course, the value of a future flooding of this tremendous magnitude in today's dollars depends on how much we value the future compared with the present—in economic terminology, the discount rate.

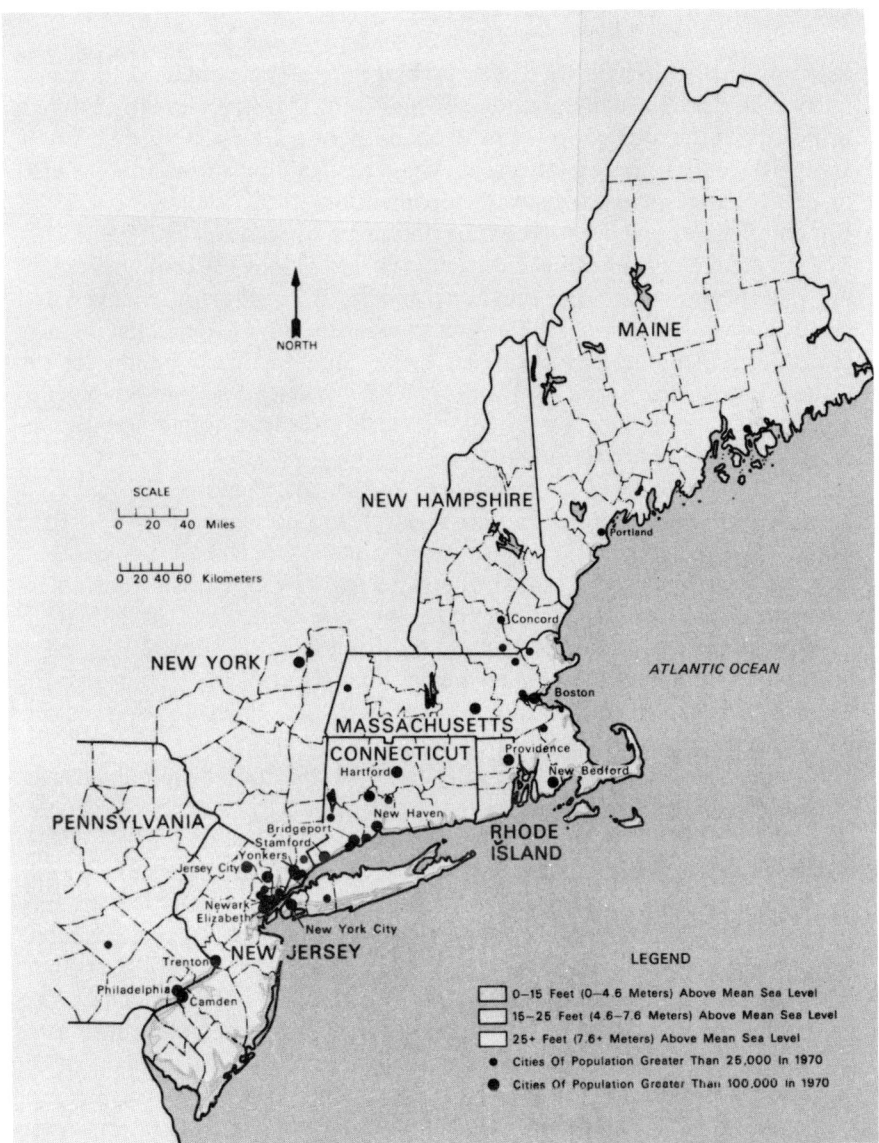

Source: S.H. Schneider and R.S. Chen, "Carbon Dioxide Warming and Coastline Flooding: Physical Factors and Climatic Impact," *Ann. Rev. Energy* 5 (1980):107-140. Reprinted with permission.

Figure 24-8. Approximate Areas Flooded by a 15-25-Foot (4.6-7.6-Meter) Rise in Mean Sea Level, North Atlantic Region

Carbon Dioxide and Risk

Risk is usually defined as some combination of probability and consequence. A risk is usually considered significant if it involves either a high probability of some loss, or some probability of a high loss. One generally takes risks only if there is also some chance of gaining substantial benefits or of avoiding some adversity. My contention, then, is that the risks of climatic changes and even of coastal flooding may well be significant. That is, even though considerable uncertainty exists all along the chain of reasoning that begins with CO_2 emissions and leads to climatic changes and perhaps coastal flooding, such a chain is plausible, and its potential impacts could be tremendous—as illustrated by the figure of $220 billion for the United States alone estimated for the coastal flooding case. Moreover, consequences are long term, essentially irreversible (at least for several generations), and often hard to quantify.

Clearly, more research could help reduce uncertainties. However, it must be recognized that to conduct research and not act entails some risk: the risk that society may be subject to more and worse climate changes than it would have if actions were taken now to reduce CO_2 emissions or otherwise ameliorate its impacts. Conversely, more research could save us from costly and perhaps unnecessary actions, if the CO_2 problem proves insignificant, or at least could help us improve the effectiveness of our actions. Thus the decision to study or act, or both, is itself a policy decision that involves the risks and benefits of research.

To put the risks of climate changes and coastal flooding in perspective, it is necessary to compare not only the magnitude and distribution of the risks, but also the magnitude and distribution of benefits and the availability of alternatives. The potential threat of chlorofluorocarbons (CFCs) to the stratospheric ozone layer provides an instructive case. On the one hand, CFCs used in aerosol cans were relatively easy to restrict in the United States because substitutes are available and the benefits of aerosol cans are fairly limited. On the other hand, CFCs like freon have been much harder to restrict, in part because they have no straightforward substitute as a refrigerant and because refrigeration is such an important component of our health, welfare, and life-style (for example, see Environmental Protection Agency 1980). Fossil fuels are to an even greater degree an integral part of our industrial economy and are likely to continue to be a major energy source in further economic development worldwide. Thus, in order to reduce fossil-fuel consumption, we must find substitute energy sources for which the ratio of marginal risk to marginal benefit is less than that for fossil-fuel burning. By *marginal* I mean the concept used in economics of an incremental change—we want to be sure that each new increment of risk brings with it an actual increment in real benefits (for example, better nutri-

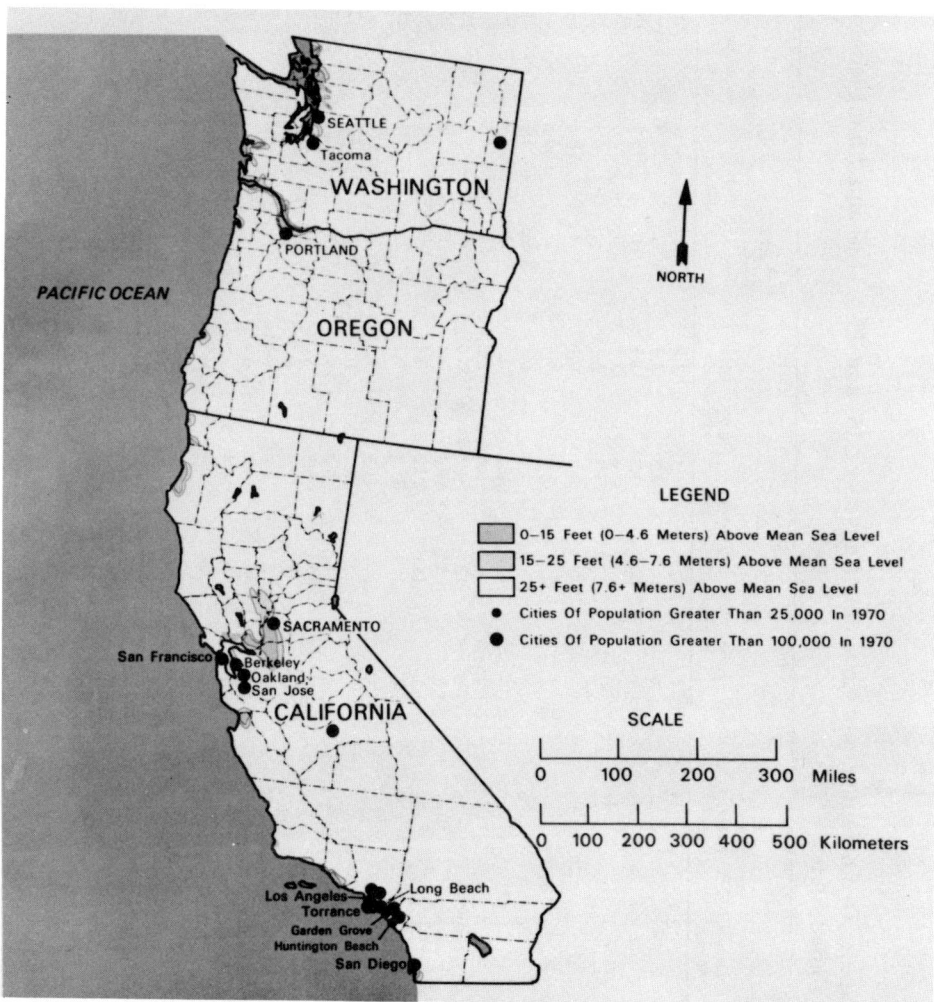

Source: S.H. Schneider and R.S. Chen, "Carbon Dioxide Warming and Coastline Flooding: Physical Factors and Climatic Impact," *Ann. Rev. Energy* 5 (1980):107-140. Reprinted with permission.

Figure 24-9. Approximate Areas Flooded by a 15-25-Foot (4.6-7.6-Meter) Rise in Mean Sea Level, West Coast Region

tion, health, and education, or increases in productive capital stock). In other words, we need to find energy sources that produce the same benefit—the same quantity and quality of useful work performed—for less risk than the risks of fossil fuels.

Table 24-2
Summary of Estimated Geographic, Demographic, and Economic Impacts of 15- and 25-ft (4.6- and 7.6-m) Rises in Sea Level for the Continental United States

		15-ft. Case				25-ft. Case		
Region/State	Percentage Flooded	Population (Millions)	Approximately Percentage of State	EMV[b] (Billions)	Approximately Percentage of State	Percentage Flooded	Population (Millions)	Approximately Percentage of State	EMV[b] (Billions)	Approximately Percentage of State

Region/State	Percentage Flooded	Population (Millions)	Approximately Percentage of State	EMV[b] (Billions)	Approximately Percentage of State	Percentage Flooded	Population (Millions)	Approximately Percentage of State	EMV[b] (Billions)	Approximately Percentage of State
Florida	24.1	2.9	43	33.4	52	35.5	3.8	55	41.7	65
Gulf Coast	4.7	2.7	—	21.3	—	5.8	3.3	—	26.9	—
Texas	2.2	0.9	8	14.3	14	3.2	1.3	11	19.2	18
Louisiana	27.5	1.7	46	6.5	51	31.4	1.8	50	7.0	55
Mississippi	1.0	c	2	0.2	3	1.5	0.1	4	0.3	5
Alabama	0.8	c	2	0.3	2	1.1	0.1	2	0.4	2
Mid-Atlantic	5.3	1.8	—	11.6	—	7.6	2.5	—	16.6	—
Georgia	2.4	0.2	4	0.8	4	4.0	0.3	6	1.2	5
South Carolina	6.7	0.3	10	1.6	12	10.2	0.3	13	2.2	16
North Carolina	7.9	0.2	3	1.4	4	9.8	0.2	4	1.9	6
Virginia	3.1	0.7	16	4.1	12	4.8	1.0	21	5.8	16
Maryland	12.3	0.2	6	1.5	5	17.3	0.4	10	2.6	9
District of Colombia	15.0	0.1	15	1.2	15	20.0	0.2	20	1.6	20
Delaware	16.0	0.1	19	1.0	18	25.0	0.1	25	1.3	25
North-Atlantic	0.9	3.6	—	33.3	—	1.3	5.0	—	47.6	—
New Jersey	9.5	0.7	9	6.2	9	13.1	0.9	12	8.4	12
Pennsylvania	0.1	0.4	3	2.0	3	0.2	0.5	4	2.8	4
New York	0.6	2.1	12	22.0	12	0.8	3.0	16	30.9	16
Connecticut	1.2	c	2	0.6	2	2.2	0.1	3	1.1	3
Rhode Island	3.5	c	3	0.2	3	7.0	0.1	6	0.4	5
Massachusetts	2.1	0.3	4	2.2	5	3.4	0.4	8	3.7	8
New Hampshire	0.1	c	0	c	1	0.4	c	1	0.1	1
Maine	0.2	c	1	0.1	2	0.4	c	1	0.2	3
West Coast	0.6	0.8	—	7.8	—	1.2	1.4	—	14.2	—
California	1.0	0.7	4	7.3	3	1.7	1.2	6	12.7	6
Oregon	0.1	c	1	0.3	1	0.4	0.1	3	0.6	3
Washington	0.4	c	1	0.2	1	7.1	0.1	2	0.9	3
All regions	—	11.6	—	107.5	—	—	15.7	—	146.9	—

| Percentage of continental U.S. | 1.5 | 5.7 | — | 6.2 | — | 2.1 | 7.8 | — | 8.4 | — |

Source: S.H. Schneider and R.S. Chen, "Carbon Dioxide Warming and Coastline Flooding: Physical Factors and Climatic Impact," *Ann. Rev. Energy* 5(1980):107-140. Reprinted with permission.

[a]Totals may not add exactly due to rounding errors.

[b]Estimated Market Value derived by dividing the "locally assessed taxable real property" in each county by the "aggregate assessment sales price ratio," which is based on a sample of market values. From Bureau of the Census, *1972 Census of Governments*, vol. 2, *Taxable Property Values and Assessment-Sales Price Ratios*, parts 1 and 2 (Washington, D.C.: U.S. Government Printing Office, 1973).

[c]Less than 0.1.

Risks do vary greatly in character, however, and different people value different risks differently. For example, developing countries may be much less concerned with relatively remote problems like CO_2-induced climate changes than are developed countries, in light of the developing nations' pressing needs for energy, food, and other resources. Comparisons between consequences measured in different terms—such as lives lost, dollars lost, and land permanently flooded—raise difficult normative issues, such as the value of life or the degree to which society as a whole values the future. These issues make risk comparisons extremely difficult and value laden (for example, see Fischhoff 1977).

However, some conclusions and comparisons can be drawn, even given our limited understanding of the risks and benefits of energy systems. For example, as several recent studies have concluded, conservation in the sense of increased end-use efficiency is certainly the least risky and most beneficial new "energy source." That is, conservation allows us to perform the same tasks with less energy and therefore fewer energy-related risks (Stobaugh and Yergin 1979a,b; Resources for the Future 1979; National Energy Strategies Project 1979; Massachusetts Energy Office 1978).

In the case of liquid fossil fuels used in automobiles, at present there are few adequate substitutes. Thus, for example, the United States seems to be willing to bear the risks of instability in the Middle East and the costs of exporting dollars there. Were the uncertain, far-off risks of climate changes incorporated into this risk/benefit decision, it is improbable that its decision would change.

In the case of electricity generation, climatic risks might well figure more prominently, since alternatives are likely to be more readily available than in transportation. However, some of these alternatives may also pose risks that may in fact be akin to the risks of climatic changes. Nuclear power, for example, involves extensive risks that are uncertain in probability and high in consequence—namely, the risks of reactor meltdowns, nuclear proliferation, and waste disposal. Like climatic changes, these could have long-term, irreversible, and global consequences that are hard to quantify. As in the case of this past summer's climatic extremes, unexpected synergisms and interactions, such as those evident in the Three Mile Island accident, may be important contributors to overall impacts. Thus, given present levels of uncertainty over the likelihood and potential consequences of both climatic changes and nuclear-power risks, the risks of climatic changes do not argue that nuclear power should be substituted for fossil fuels in the generation of electricity. To do so would only substitute nuclear-power risks of a similar nature and possibly of equivalent magnitude to that of climatic risks. Instead, to the degree possible, we should look for energy alternatives that are not plagued with uncertain, large, and irreversible risks.

The risk of CO_2-induced climatic warming is but one factor among many to be considered in making energy choices. But this risk, like the risks of some other energy sources, has important ramifications from a long-term, global, and indeed ethical perspective. Can we take the risk of a future climate as different from the present as, say, the present is from the last Ice Age? Can we expect to improve the lot of the world's population through intensive development without destroying the earth's ability to maintain a stable, relatively benign environment? What legacy—of flooded lands, deforested and eroded countrysides, dangerously radioactive wastelands, or perhaps even a world destroyed by nuclear war—can we expect to leave unless we look squarely into the future? This, I submit, is the message of the carbon dioxide problem: it forces us to deal imaginatively, cooperatively, and with as much foresight as we can muster with the future of the human race and the planet on which all of us live. To quote once more from an academy study (Climate Research Board 1980), "this earth is the only planet we have on which to live." It is our only home and only hope.

References

Adams, J.A.S.; Mantovani, M.S.M.; and Lundell, L.L. 1977. "Wood Versus Fossil Fuel as a Source of Excess Carbon Dioxide in the Atmosphere: A Preliminary Report." *Science* 196:54-56.

Aharon, P.; Chappell, J.; and Compston, W. 1980. "Stable Isotope and Sea-Level Data from New Guinea Supports Antartic Ice-Surge Theory of Ice Ages." *Nature* 283:649-651.

Arrhenius, S. 1896. "On the Influence of Carbonic Acid in the Air upon the Temperature of the Ground." *Philos. Mag.* 41:237-276.

Ausubel, J. 1980. "Economics in the Air—An Introduction to Economic Issues of the Atmosphere and Climate." In *Climatic Constraints and Human Activities,* J. Ausubel and A.K. Biswas, eds. New York: Pergamon, pp. 13-59.

Bacastow, R.B. 1976. "Modulation of Atmospheric Carbon Dioxide by the Southern Oscillation." *Nature* 261:116-118.

———. 1977. "Influence of the Southern Oscillation on Atmospheric Carbon Dioxide." In *Fate of Fossil Fuel CO_2 in the Oceans,* N.R. Anderson and A. Malahoff, eds. New York: Plenum, pp. 33-43.

Bolin, B. 1977. "Changes of Land Biota and Their Importance for the Carbon Cycle," *Science* 196:613-615.

Bolin, B., and Bischof, W. 1970. "Variations of the Carbon Dioxide Content of the Atmosphere in the Northern Hemisphere." *Tellus* 22:431-442.

Bolin, B., and Keeling, C.D. 1963. "Large-Scale Atmospheric Mixing as Deduced from the Seasonal and Meridional Variations of Carbon Dioxide." *J. Geophys. Res.* 68:3899-3920.

Bureau of the Census. 1973a. *County and City Data Book, 1972, A Statistical Abstract Supplement.* Washington, D.C.: U.S. Government Printing Office.

_____. 1973b. *1972 Census of Governments, vol. 2, Taxable Property Values and Assessment-Sales Price Ratios, parts 1 and 2.* Washington, D.C.: U.S. Government Printing Office.

Callendar, G.S. 1958. "On the Amount of Carbon Dioxide in the Atmosphere." *Tellus* 10:243-248.

Carbon Dioxide Effects Research and Assessment Program. 1980. *Workshop on Environmental and Societal Consequences of a Possible CO_2-Induced Climate Change. Report 009, CONF-7904143.* National Technical Information Service, Springfield, Va.

Center for Environmental Assessment Services. 1980. *Impact Assessments of Major Climatic and Other Natural Events.* Monthly report issued by the National Oceanic and Atmospheric Administration, U.S. Department of Commerce, Washington, D.C.

Chen, R.S. 1980. "Impacts of Carbon Dioxide Induced Climate Change." *Proceedings: Bio-Energy '80, World Congress and Exposition.* Washington, D.C.: Bio-Energy Council, pp. 544-547.

Climate Research Board. 1979. *Carbon Dioxide and Climate: A Scientific Assessment.* Washington, D.C.: National Academy of Sciences.

_____. 1980. "Letter Report of the *Ad Hoc* Study Panel on Economic and Social Aspects of Carbon Dioxide Increase." Washington, D.C.: National Academy of Sciences.

Denton, G.H.; Armstrong, R.L.; and Stuiver, M. 1971. "The Late Cenozoic Glacial History of Antarctica." In *The Late Cenozoic Glacial Ages,* K. Turekian, ed. New Haven: Yale University Press, pp. 267-306.

Environmental Protection Agency. 1980. "Ozone-Depleting Chlorofluorocarbons: Proposed Production Restriction." *Federal Register* 45:66726—66734 (Tuesday, 7 October 1980; 40 CFR, part 762).

Fischhoff, B. 1977. "Cost Benefit Analysis and the Art of Motorcycle Maintenance." *Policy Sciences* 8:177-202.

Hock, J. 1980. "Presentation to the National Climate Program Advisory Committee." Washington, D.C.: U.S. Department of Commerce (27 October).

Hollin, J.T. 1969. "Ice Sheet Surges and the Geological Record." *Can. J. Earth Sci.* 6:911-918.

_____. 1972. "Interglacial Climates and Antarctic Ice Surges." *Quat. Res.* 4:385-404.

_____. 1980. "Climate and Sea Level in Isotope Stage 5: An East Antarctic Ice Surge at 95,000 BP?" *Nature* 283:629-633.

Hughes, T. 1973. "Is the West Antarctic Ice Sheet Disintegrating?" *J. Geophys. Res.* 78:7884-7910.

Jones, P.D., and Wigley, T.M.L. 1980. "Northern Hemisphere Temperatures 1881-1979." *Climate Monitor* 9:43-47.

Keeling, C.D. 1978. "Atmospheric Carbon Dioxide in the 19th Century (Letter)." *Science* 202:1109.

Keeling, C.D.; Adams, J.A., Jr.; Ekdahl, C.A., Jr.; and Guenther, P.R. 1976a. "Atmospheric Carbon Dioxide Variations at the South Pole." *Tellus* 28:552-564.

Keeling, C.D., and Bacastow, R.B. 1977. "Impact of Industrial Gases on Climate." In *Energy and Climate,* Geophysics Study Committee. Washington, D.C.: National Academy of Sciences, pp. 72-95.

Keeling, C.D.; Bacastow, R.B.; Bainbridge, A.E.; Ekdahl, C.A., Jr.; Guenther, P.R.; and Waterman, L.S. 1976b. "Atmospheric Carbon Dioxide Variations at Mauna Loa Observatory, Hawaii." *Tellus* 28: 538-551.

Kellogg, W.W., and Schware, R. 1980. *Climate Change and Society: Consequences of Increasing Carbon Dioxide.* Boulder, Colo.: Aspen Institute for Humanistic Studies and Westview Press.

MacDonald, G.J.F. 1978. *An Overview of the Impact of Carbon Dioxide on Climate.* McLean, Va.: Mitre Corporation.

MacRae, J.E., and Graedel, T.E. 1979. "Carbon Dioxide in the Urban Atmosphere: Dependencies and Trends." *J. Geophys. Res.* 84:5011-5017.

Manabe, S., and Wetherald, R.T. 1975. "The Effects of Doubling the CO_2 Concentration on the Climate of a General Circulation Model." *J. Atmos. Sci.* 32:3-15.

Marland, G., and Rotty, R.M. 1979. "Carbon Dioxide and Climate." *Rev. Geophys. Space Phys.* 17:1813-1824.

_____. 1980. "Que faire devant le degagement croissant de CO_2 dans l'atmosphere?" *Consensus* 2-3:41-52.

Massachusetts Energy Office. 1978. *New England Energy Policy Alternatives Study.* Boston: Massachusetts Energy Office.

Mercer, J.H. 1968. "Antarctic Ice and Sangamon Sea Level." In *Int. Assoc. Sci. Hydrol.,* Commission of Snow and Ice, General Assembly of Bern, Publ. no. 79, pp. 217-225.

_____. 1978. "West Antarctic Ice Sheet and CO_2 Greenhouse Effects: A Threat of Disaster." *Nature* 271:321-325.

Mitchell, J.M. Jr. 1977. "The Changing Climate." In *Energy and Climate,* Geophysics Study Committee. Washington, D.C.: National Academy of Sciences, pp. 51-58.

National Energy Strategies Project. 1979. *Energy in America's Future.* Baltimore, Md.: Johns Hopkins University Press.

Neumann, A.C., and Moore, W.S. 1975. "Sea Level Events and Pleistocene Coral Ages in the Northern Bahamas." *Quat. Res.* 5:215-224.

Paltridge, G.W., and Platt, C.M.R. 1976. *Radiative Processes in Meteorology and Climatology, Developments in Atmospheric Science*, vol. 5. New York: Elsevier.

Panel IV. 1980. "Social and Institutional Responses." In *Workshop on Environmental and Societal Consequences of a Possible CO2-Induced Climate Change*, Carbon Dioxide Effects Research and Assessment Program, Report 009, CONF-7904143. Springfield, Va.: National Technical Information Service, pp. 79-103.

Pearman, G.I. 1977. "Further Studies of the Comparability of Baseline Atmospheric Carbon Dioxide Measurements." *Tellus* 29:171-180.

Perry, H., and Landsberg, H.H. 1977. "Projected World Energy Consumption." In *Energy and Climate*, Geophysics Study Committee. Washington, D.C.: National Academy of Sciences, pp. 35-50.

Resources for the Future. 1979. *Energy: The Next Twenty Years*. Cambridge, Mass.: Ballinger.

Rotty, R.M. 1979. "Present and Future Production of CO_2 from Fossil Fuels—A Global Appraisal." In *Workshop on the Global Effects of Carbon Dioxide from Fossil Fuels*, Carbon Dioxide Effects Research and Assessment Program, Report 1, CONF-770385, Springfield, Va.: National Technical Information Service, pp. 36-43.

Rust, B.W.; Rotty, R.M.; and Marland, G. 1979. "Inferences Drawn from Atmospheric CO_2 Data." *J. Geophys. Res.* 84:3115-3122.

Schneider, S.H. 1975. "On the Carbon Dioxide-Climate Confusion." *J. Atmos. Sci.* 32:2060-2066.

Schneider, S.H., and Chen, R.S. 1980. "Carbon Dioxide Warming and Coastline Flooding: Physical Factors and Climatic Impact." *Ann. Rev. Energy* 5:107-140.

Schneider, S.H., with Mesirow, L.E. 1977. *The Genesis Strategy: Climate and Global Survival*. New York: Delta.

Schneider, S.H.; Washington, W.M.; Chervin, R.M. 1978. "Cloudiness as a Climatic Feedback Mechanism: Effects on Cloud Amounts of Prescribed Global and Regional Surface Temperature Changes in the NCAR GCM." *J. Atmos. Sci.* 35:2207-2221.

Stobaugh, R., and Yergin, D. 1979a. "After the Second Shock: Pragmatic Energy Strategies." *Foreign Affairs* 57:836-871.

_____, eds. 1979b. *Energy Future: Report of the Energy Project at the Harvard Business School*. New York: Random House.

Stuiver, M. 1978a. "Atmospheric Carbon Dioxide and Carbon Reservoir Changes." *Science* 199:253-258.

_____. 1978b. "Atmospheric Carbon Dioxide in the 19th Century (Reply to Keeling, 1978)." *Science* 202:1109.

Sundquist, E.T., and Miller, G.A. 1980. "Oil Shales and Carbon Dioxide." *Science* 208:740-741.

Thomas, R.H.; Sanderson, T.J.O.; and Rose, K.E. 1979. "Effect of Climatic Warming on the West Antarctic Ice Sheet." *Nature* 277:355-358.

Thompson, S.L., and Schneider, S.H. 1979. "A Seasonal Zonal Energy Balance Climate Model with an Interactive Lower Layer." *J. Geophys. Res.* 84:2401-2414.

White, G.F., and Haas, J.E. 1975. *Assessment of Research on Natural Hazards*. Cambridge, Mass.: MIT Press.

Wilson, A.T. 1969. "The Climatic Effects of Large-Scale Surges of Ice Sheets. *Can. J. Earth Sci.* 6:911-918.

Woodwell, G.M.; Whittaker, R.H.; Reiners, W.A.; Likens, G.E.; Delwiche, C.C.; and Botkin, D.B. 1978. "The Biota and the World Carbon Budget." *Science* 199:141-146.

Workshop on Alternative Energy Strategies. 1977. *Energy: Global Prospects 1985-2000*. New York: McGraw-Hill.

World Energy Conference. 1978. *World Energy Resources 1985-2020: Coal Resources*. Guildford, England: IPC Press.

25 Risk-Assessment Analysis for Various Energy Sources: Some Annotations

Klaus Gottstein

Standards for Risks

Every new technology produces its own new risks. That was so with the introduction of electricity into households, with automobiles, with air traffic, with X-ray diagnosis. As time goes by, more experience is collected about these risks, and standards of safety are established and prescribed by law. Nuclear energy is now reaching this phase. Safety standards are already rather well developed, although still subject to improvement, with respect to the operation of nuclear reactors. With respect to the other parts of the fuel cycle, however, much work remains to be done. It will be necessary to set criteria for geological waste disposal, spent-fuel storage, abandoned uranium mines, and tailings from mines and mills. Although the maximum hazard resulting from inadequate waste disposal is much smaller than that which could be postulated as the result of a reactor accident or sabotage, performance and site criteria for geological waste disposals should, as the CONAES report states, be set immediately (leach rates, heat rates, groundwater standards, seismic stability standards, drilling restrictions).[1]

Measures to reduce risk usually cost money and thereby reduce the commercial competitiveness of those introducing them. In order to avoid a situation in which risk consciousness is punished, international agreements on safety standards are required in all relevant fields. This is also true for the development of synthetic-fuel industries as an insurance against the risk of political disruptions in oil-producing regions, at a time when such industries are not yet commercially viable.

Risks of Various Energy Sources

All energy systems entail risks. Coal is often said (see, for example, the CONAES report) to be ten times more dangerous than any other type of

---

This chapter was first published in *Atomkernenergie/kerntechnik* 37, no. 4 (1981):241-243. Reprinted with permission. These annotations were made after reading the CONAES report. Several of them are quotations from the report. Others are modifications of the statements in that report. Still others came from other sources or are the sole responsibility of the author, who is also responsible for the selection and arrangement of the statements here compiled.

energy. On the other hand, the risk ranges of coal and nuclear energy overlap if the most optimistic limit of coal is compared with the most pessimistic limit of nuclear energy. The risks to health, climate, water supply, and the ecology in general will probably limit the use of coal at three times the current output. However, the burning of coal is not the only source of air pollution. Automobiles make an important contribution. Of particular concern are climatic risks: a redistribution of agricultural productivity caused by climatic changes from CO_2 production could have disastrous consequences for some regions. Global effects could also result from a widespread use of solar energy by changes in the surface reflectivity of the earth and/or temperature changes of tropical oceans. Energy parks of more than 30 GW_e, as they are sometimes suggested for security and efficiency reasons, could change regional circulation and have climatic effects. Ecological problems must also be expected from large-scale hydroelectric power projects and from large biomass farms.

The Rasmussen Report (WASH-1400, October 1975) in chapter 7 of the main report and the report of the American Physical Society (APS) study group on light-water-reactor safety (*Reviews of Modern Physics* 47, supp. 1, 1975) list a number of (solvable) problems that need further investigation. It would be advisable to distribute reports on progress in these investigations as widely as WASH-1400 and the APS study were distributed.

Risks, Costs, Psychology, Philosophical Values

Risks for various energy sources can be compared only if their probability distributions are known well enough so that statements can be made like "per gigawatt-year used from energy source X_i, there is a probability y_{ij} that j persons suffer physical injuries or discomfort from causes germane to this energy source, with k_1 immediate fatalities, k_2 premature but delayed deaths, k_3 injuries and diseases of other types, k_4 cases of damage to the general quality of life, with $k_1 + k_2 + k_3 + k_4 = j$."

In practice it will be very difficult to determine these probabilities with any precision, not only because for different scenarios the relation between k_1, k_2, k_3, and k_4 could vary considerably, but also because there are just too many parameters involved. Our knowledge, our forecasting capabilities, and our analytical capacity are insufficient for dealing with this problem accurately. We can only hope to get approximate estimates of these probabilities and of the associated error bands. For example, such estimates seem to show that solar energy, per kilowatt-hour produced, will involve much higher risks for accidents than coal, oil, or nuclear energy because of the large quantities of material required and the huge industrial efforts needed in mining, refining, fabricating, and constructing collectors, storage systems, and related apparatus.

So far we are speaking only about personal risks. But there are also risks to the environment and in particular to the climate that must be taken into account. There is also our esthetic sense, which reacts differently to various forms of energy use. The relative weight that should be given to these environmental and esthetic considerations is a matter of one's philosophical values. This is also true for the question of the price society will be willing to pay for a "sufficient" supply of energy. This price has an economic component, a health component, an esthetic component. The definition of what is sufficient can also be made only on the basis of the philosophical values that manifest themselves in the political process. A detailed analysis of values and motivations in Western industrialized societies might well show that individuals have unconsciously become used to the concept of the welfare state—the concept that the state is responsible for the widest possible removal of all risks from life as well as for an equal distribution of the benefits of life. On the other hand, it seems that many people do not realize that such a welfare state is only possible, first, under conditions of a high standard of living, which in turn depends on a plentiful energy supply, which in turn has a price, as mentioned before, or, second, under conditions of dictatorship, where nearly everything, including risks and benefits, is distributed by the state.

Even such basic achievements as equal rights for women most probably would not have been made in our time had energy-intensive technical progress (washing machine, vacuum cleaner, refrigerator, dishwasher, central heating, electric stove, and so forth) not arrived to free women from the labor that had occupied their grandmothers from dawn to dusk. Their grandfathers, likewise, worked sixteen hours a day to make a meager living on farms or in factories without the energy-consuming tools of today.

Dictatorship will be ruled out by most people because its price in terms of loss of freedom is too high. Welfare for our neighbors and ourselves is certainly desirable, and much can be said for a welfare state with a high standard of living. On the other hand, we should have no illusions about its price in terms of industrialization, energy consumption, and bureaucracy.

It is often said that the risks connected with nuclear energy should be compared with the risks connected with a general shortage of energy. This is certainly true. It is also often said that a general shortage of energy might lead to war for the possession of the remaining resources. Whereas this cannot be ruled out completely, the opposite might also be true. Shortage of energy might force nations to agreements and cooperation because war, apart from being energy intensive, might endanger the remaining resources.

Whereas there is much discussion of the risks resulting *from* the use of various energy sources, little attention is paid to the problems of risks *to* energy supply from considerations of environmental protection, public health, national security, economic growth, and equity among regions and classes. All these considerations have their price, not only in money but also in terms of energy consumption, a price that may or may not seem appropriate from the point of view of the value system applied.

The acceptability of risks is determined not only by their magnitude, by economic factors, and by the value system of those who must accept the risk, but also by psychological factors. For example, one often reads the statement that the populations of the Western countries accept without difficulty tens of thousands of traffic deaths year after year, whereas many people become excited about the very remote possibility of a large reactor accident killing a few hundred people. Although it is certainly true that, for psychological reasons, one major accident gets more attention than several small ones with the same total number of casualties—if the 16,000 or so people killed by traffic accidents in the Federal Republic of Germany annually would all die together on one single day, our present traffic regulation would probably no longer be acceptable, even though the total number of deaths per year would not be different—one should not forget that there are also objective differences between the two cases: our social system with its ambulances, hospitals, nurses, and cemeteries is adapted to 16,000 traffic deaths spread over 365 days but would break down if all these deaths occured on one day. This breakdown would affect the general public (through traffic jams, unavailability of hospitals and doctors to normal patients, and so on) so that one major accident really is much worse than many small ones with the same total death toll.

The problem is that there are no accepted common standards for comparing different kinds of risks. In principle, there are three different bases for the comparison of risks from energy sources:

1. Compare the risks from a particular energy source with background risks. (*Example*: Compare the risk of getting cancer as a consequence of the existence of a nuclear power plant with the general risk of getting cancer; or compare the radiation from a nuclear power plant with natural radiation.)
2. Compare alternative energy risks. (*Example*: Compare risks from coal or hydroelectric power against risks from nuclear energy.)
3. Compare energy risks against risks from other technologies. (*Example*: Compare nuclear fatalities with fatalities from air traffic.)

All these methods of comparison have the following peculiarities in common:
1. It makes a great psychological difference to public perception whether comparisons are made in percentages or in absolute numbers.
2. There is no objective, quantitative way to relate injury and sickness to fatalities or immediate death to death in future populations. These

relations are subject to value judgments. Values may change with time.
3. Familiar risks tend to be accepted more easily than new, unfamiliar risks.
4. High, voluntary risks are accepted more easily than small, involuntary risks.

In general, delayed risks (like those of cigarette smoking) are more acceptable to the public than immediate risks (like allowing hunting near homes). However, with nuclear risks the general feeling seems to be the other way around. This is probably true because radiation is not yet well understood by the general public, so that its delayed effects create an uneasy feeling.

Regulations like the Clean Air Act and the Water Quality Act adopt scientific standards of safety with no consideration of cost and feasibility. Because this often means that compliance would be excessively expensive or would even require the closing down of plants important for the economy of a region, it is often not enforced. The distribution of risks and benefits should be studied from a sociopolitical and socioeconomic point of view in order to make regulations enforceable. On the other hand, economic factors should not be overestimated. Public attitudes have symoblic dimensions. Nuclear energy, for example, stands for big business and big government unresponsive to local needs. Solar energy, on the other hand, stands for natural forms of energy that can be locally controlled, regardless of the high risks that would be connected with a widespread use of solar energy. On the other hand, again, one can hear the opinion that mandated conservation would mean an intrusion of government into the private life of the citizen.

On the other hand, there is a widespread feeling today that there are also secondary risks connected with energy affluence. If energy is plentiful and cheap, industrial production and human activities in general proliferate, with concomitant intrusion on nature. It is again a question of value judgment to what extent this is desirable or tolerable, and where the limits should be.

Social and psychological studies should also be undertaken regarding the social risks of alienation by centralized technologies that citizens cannot understand or control (with the possible consequence of crime and terrorism, drug abuse, turning away of students from technical disciplines, and erosion of civil liberties and democratic institutions).

Because of the importance of public perception of risks, information on risk assessment, its value dependence, and cost-benefit analysis should be made available in a form understandable to the average citizen. The magnitude of risks in general depends on details that cannot be foreseen reliably before full-scale design and operation.

Political Risks

Risks resulting from energy shortages and from the vulnerability of different energy regimes have been insufficiently studied. Such studies should be stepped up and the results disseminated to the public.

It is a risk in itself that concentration on energy risks diverts attention from the quite general risks of war, poverty, disease, crime, and "normal" accidents. On the other hand, it is perhaps understandable that one tends to concentrate first on those risks for which measures for their reduction seem visible. It is much more difficult to do something about war, poverty, and so forth. Also, it is not at all certain that the problems of war and poverty would be attacked more vigorously if energy risks were to be neglected. It is a serious consideration, however, that the other risks could increase as a consequence of a scarcity of energy.

The CONAES report states that only alternative energy risks should be compared (page 50). Some committee members disagree, insisting that all risks should be compared. It seems that both statements are true, but for different groups of people. Political decision makers, who have to set general priorities, must look at all risks. Energy planners, who have no business with war, disease, or airline accidents, can only look at alternative energy risks. This is the first step. Their results on relative energy risks and on costs for risk reduction must enter into the overall political calculation of risks within the more complex systems of which energy systems represent only one dimension.

In order to reduce the risk of social and political disruptions, the transition from gradually depleting resources of oil and gas to new technologies should be made as smooth and orderly as possible.

WASH-1400 does not investigate the contributions to risk by war and sabotage. Since war and sabotage cannot be ruled out, the attempt should be made to assess these risks for countries that expect to rely heavily on the production of electricity by nuclear power. It could be that the risk of war breaking out would be reduced by the presence of considerable numbers of nuclear reactors in the countries concerned (as seems to be the case between countries possessing nuclear weapons) because of the necessity to avoid total destruction of these reactors at all costs. Fetter and Tsipis (Report no. 5, Program in Science and Technology for International Security, Department of Physics, Massachusetts Institute of Technology, September 1980) have calculated the devastating effects of the worst case of a nuclear weapon exploding on a nuclear reactor.

Proliferation is mainly a political risk, not a technical one. It can be avoided only by removing the incentives for nuclear armaments.

Note

1. "Energy in Transition 1985-2010," Final Report of the Committee on Nuclear and Alternative Energy Systems (CONAES), National Research Council, National Academy of Sciences, San Francisco: W.H. Freeman and Company, 1979.

Appendix: Program

Forum Moderators and Organizers

Behram Kursunoglu (forum chairman), University of Miami
Marcelo Alonso, Florida Institute of Technology
Hans A. Bethe, Cornell University
Bernard Chadenet, The World Bank, Washington, D.C.
Karl Cohen, Stanford University
Jean Couture, Societe Generale, Paris
Dante B. Fascell, Member, U.S. Congress
Anthony J. Favale, Grumman Aerospace Corporation, New York
Don Fuqua, Member, U.S. Congress
Pierre Grau, Framatome, France
Robert Hofstadter, Stanford University
Mike McCormack, Member, U.S. Congress
A.J. Meyer II, The Chase Manhattan Bank, N.A., New York
Louis Neel, Universite de Grenoble, France
David Nissen, The Chase Manhattan Bank, N.A. New York
Carl D. Pursell, Member, U.S. Congress
Edward Teller, Hoover Institution; Stanford University
Victor Urquidi, El Colegio de Mexico
Eugene Wigner, Princeton University
I.G.K. Williams, OECD, Paris
C. Pierre L. Zaleski, Embassy of France, Washington, D.C.
Andrew C. Millunzi (forum editor), U.S. Department of Energy
Arnold Perlmutter (forum editor), University of Miami
Linda Scott (secretary of the forum), University of Miami
Helga S. Billings (forum secretariat), University of Miami

Sunday, 9 November 1980

    12:00 Noon   Registration and Forum Check-In Information
                           (free day for participants)

     5:30 p.m.   Meeting of the Scientific Council of the Center for Theoretical
                           Studies (2nd floor, Board Room East)

                       Members of the Scientific Council:

                       Karl Cohen, Stanford University
                       Manfred Eigen, Max-Planck-Institute für Biophys. Chemie,
                           West Germany
                       Robert Hofstadter, Stanford University

Behram Kursunoglu, chairman, CTS, University of Miami
Willis E. Lamb, Jr., University of Arizona
Arnold Perlmutter, secretary, CTS, University of Miami
Julian Schwinger, University of California, Los Angeles
Alfred Sklar, Miami, Florida
Edward Teller, Hoover Institution, Stanford University
Eugene Wigner, Princeton University

7:30 p.m.   *Dinner for Members of the Scientific Council* (by invitation)
Hosts: Mr. and Mrs. Grisha Hovsebian, Fort Lauderdale, Florida

Monday, 10 November 1980, Seaview Room

8:45 a.m.   Prologue

9:00 a.m.   *Session I*
Energy in the Critical Decade of the 1980s

Moderator: *Behram Kursunoglu*, University of Miami

Dissertators: *Jean Claude Balaceanu*, French Petroleum Institute, on behalf of *Andre Giraud*, Minister of Industry, France, "French Politic for Energy during the Coming Decade"
Mike McCormack, Member, U.S. Congress, "The Urgent Need for Realistic Planning Now"

10:15-10:30 a.m.   *Coffee Break*

M.E.J. O'Loughlin, Exxon Corporation, New York, "Getting from Here to There: The Energy Road Ahead"
Edward Teller, Hoover Institution, Stanford University, "Geopolitics of the Persian Gulf"

Annotators: *Kjell Hakansson*, Studsvik Energiteknik AB, Sweden
Joseph Kestin, Brown University
C. Pierre L. Zaleski, Embassy of France, Washington, D.C.
John Savoy, Sun Company, Pennsylvania

12:15-1:45 p.m.   *Lunch Break*

1:45 p.m.   *Session II*
Energy in the Critical Decade of the 1980s (continued)

Moderator: *Karl Cohen*, Stanford University

Dissertators: *Wolfgang Sassin*, IIASA, Austria, "Towards a New Order for a Global Energy System"
Sam H. Schurr, EPRI, California, "International Aspects of America's Energy Choices"

Appendix

3:00-3:15 p.m.	*Coffee Break*

Richard H. Sheehan, The World Bank, Washington, D.C., "Energy in the Developing Countries"

Eugene Wigner, Princeton University, "Nuclear Fuel Reprocessing and Waste Disposal—A Universal Problem"

Annotators: *Therese deMazancourt*, Electricite de France
Edmund P. Gaines, Vermont Yankee Nuclear Power Corporation
L.G. Hauser, Westinghouse Electric Corporation, Pennsylvania
A.J. Meyer II, The Chase Manhattan Bank, N.A., New York

5:00 p.m.	Forum Adjourns for the Day
6:00-7:00 p.m.	*Cocktail Reception*, courtesy of the SunRise Inn (Penthouse Floor—Starlight Room)
9:00 p.m.	*Evening Lecture* Seaview Room

Manfred Eigen, Max-Planck-Institut für Biophys. Chemie, West Germany, "Evolution of the Genetic Code"

Tuesday, 11 November 1980

9:00 a.m.	*Session III* Energy and Economic Policy in Developed Economies (including substitution of fuels)
Moderator:	*David Nissen*, The Chase Manhattan Bank, N.A., New York
Dissertators:	*Leslie Grainger*, National Coal Board, United Kingdom, "The Role of Coal"
	L.G. Hauser, Westinghouse Electric Corporation, Pennsylvania, "The Role of Electricity as a Substitute for Liquid Fuels"
10:15-10:30 a.m.	*Coffee Break*

N.P. Kannan, MITRE Corporation, Virginia, "A Disequiliblibrium Effect of Energy Price Shocks"

Thomas H. Lee, General Electric Company, Connecticut, "Role of Electricity in Solving the Energy Problems in the Near Term"

Annotators:	*W.G. Jensen*, National Coal Board, United Kingdom *Chihiro Kikuchi*, University of Michigan *S.S. Penner*, University of California at San Diego *Michael L. Telson*, U.S. House of Representatives
12:15-1:45 p.m.	*Lunch Break*
1:45 p.m.	*Session IV* Oil Production Policy and Economic and Social Development in Oil-Exporting Countries
Moderator:	*Victor L. Urquidi*, El Colegio de Mexico
Dissertators:	*Abel Beltran del Rio*, Wharton EFA, Pennsylvania, "Mexican Oil Policy Macroeconomics in the Past"
3:00-3:15 p.m.	*Coffee Break*
	Panel Discussion
Annotators:	*Mariano Bauer*, Universidad Nacional Autonoma de Mexico *Charles F. Cook*, Phillips Petroleum Company, Oklahoma *Tania Loynaz De Moron*, Ministry of Energy and Mines, Venezuela *C.H. Reing*, Mobil Research & Development, New York
5:00 p.m.	*Forum Adjourns for the Day*

Wednesday, 12 November 1980

9:00 a.m.	*Session V* Energy Imports on Economics and Political Developments in Oil-Importing Developing Countries
Moderators:	*Marcelo Alonso*, Florida Institute of Technology *Behram Kursunoglu*, University of Miami
Dissertators:	*R.K. Pachauri*, Administrative Staff College of India, "The Political Economy of Higher Oil Import Bills in the Third World" *Martin Scholl*, MITRE Corporation, Virginia, "Energy and Development Indicators for Developing Countries"
10:15-10:30 a.m.	*Coffee Break*
	Panel Discussion
Annotators:	*Sadik Kakac*, Middle East Technical University, Turkey *Antonio Agapito Rodriguez*, Ministry of Energy and Mines, Venezuela *Richard H. Sheehan*, The World Bank, Washington, D.C.
12:15-1:45 p.m.	*Lunch Break*

1:45 p.m.	*Session VI* Energy for Developing Countries
Moderators:	*Marcelo Alonso*, Florida Institute of Technology *A.J. Meyer II*, The Chase Manhattan Bank, N.A., New York
Dissertators:	*Marcelo Alonso*, Florida Institute of Technology, "Relationship between Energy Growth and Economic Growth: The Point of View of the Developing Countries" *Eduardo Lopez-Ballori*, Office of Energy of Puerto Rico, "Alternative Energy Options for Puerto Rico"
3:00-3:15 p.m.	*Coffee Break*
	Ulises Ramirez Olmos, Ministry of Energy and Mines, Venezuela, Title to be announced
Annotators:	*Charles R. Blitzer*, International Development Corporation Agency *P.F. Castellon*, Kaiser Aluminum & Chemical Corporation, California *Fernando Prieto Calderon*, Universidad Nacional Autonoma de Mexico *Kenneth G. Soderstrom*, Center for Energy & Environment Research, Puerto Rico
5:00 p.m.	Forum Adjourns for the Day
7:00 p.m.	*Forum Banquet* (SunRise Inn) Penthouse Floor-Starlight Room

Thursday, 13 November 1980

9:00 a.m.	*Session VII* Nuclear Fusion and the Long-Term Geopolitics of Energy
Moderators:	*Anthony J. Favale*, Grumman Aerospace Corporation, New York *Robert Hofstadter*, Stanford University
Dissertators:	*Bruno Coppi*, Massachusetts Institute of Technology, "Near-Term Feasibility of the D-He$_3$ Reactor" *Gunther Grieger*, Max-Planck-Institut für Plasmaphysik, West Germany, "Stellarators for Steady State Fusion Reactors"
10:15-10:30 a.m.	*Coffee Break*

	John F. Holzrichter, Lawrence Livermore Laboratory, "Inertial Fusion and Energy Production" *N. Douglas Pewitt*, U.S. Department of Energy, "Nuclear Fusion and the U.S. Energy Policy" *Paul Reardon*, Princeton Plasma Physics Laboratory, "Status of Tokamak Development in the United States"
Annotators:	*Ira Bornstein*, Argonne National Laboratory *R.H. Davis*, Florida State University *Roy R. Johnson*, KMS Fusion, Inc., Michigan *Frederick Tappert*, University of Miami *Daniel Wells*, University of Miami
12:15-1:45 p.m.	*Lunch Break*
1:45 p.m.	*Session VIII* Nuclear Fission as a World Energy Source
Moderators:	*Edward Teller*, Hoover Institution, Stanford University *Eugene Wigner*, Princeton University
Dissertators:	*R.J. Creagan*, Westinghouse Electric Corporation, Pennsylvania, "Nuclear Fission as a World Energy Source" *Peter Fortescue*, General Atomic Company, California, "Nuclear Energy—The Next Phase: Approaches to Restoration of Confidence and Motivation"
3:00-3:15 p.m.	*Coffee Break*
	William H. Hannum, OECD Nuclear Energy Agency, France "How Available Is the Nuclear Option" *Georges Vendryes*, CEA, France, "The French Case"
Annotators:	*Herbert G. Duggan*, Union Carbide Corporation, Tennessee *Henri Jammet*, CEA, France *W. Bennett Lewis*, Queen's College, Canada *Miro Todorovich*, CUNY-BCC, New York *Huseyin Yilmaz*, Winchester, Massachusetts
5:00 p.m.	Forum Adjourns for the Day

Friday, 14 November 1980

9:00 a.m.	*Session IX* Safety and Public Acceptance of Nuclear Energy
Moderator:	*Robert J. Budnitz*, Teknekron Research, Inc., California
Dissertators:	*Robert J. Budnitz*, Teknekron Research, Inc., California "The Role of Safety Research" *Saul Levine*, N.U.S. Corporation, Maryland, "The Industry Viewpoint"

Appendix

10:15-10:30 a.m.	*Coffee Break*
	Thomas Murley, U.S. Nuclear Regulatory Commission, "The Regulatory Viewpoint"
Annotators:	*R.W. DeVane, Jr.*, Combustion Engineering, Inc., Connecticut
	Ersel A. Evans, Westinghouse Hanford Company, Washington
	Lars Nojd, Stuksvik Energiteknik AB, Sweden
	Jean Pellerin, CEA, France
12:15-1:45 p.m.	*Lunch Break*
1:45 p.m.	*Session X* Risk Assessment Analysis for Various Energy Sources
Moderator:	*Pierre Grau*, Framatome, France
Dissertators:	*Philippe Hubert*, Paris, France, "Comparison of the Health Effect of Energy Changes and Assessment for the French Case"
	Wladimir Paskievici, University of Montreal, Canada, "Health Risks Associated With Electric Power Production: A Comparative Study"
	Robert S. Chen, National Academy of Sciences, Washington, D.C. "The Implications of Energy Production for Atmospheric Carbon Dioxide and Global Climate"
3:00-3:15 p.m.	*Coffee Break*
	Panel Discussion
Annotators:	*Klaus Gottstein*, Max-Planck-Institut für Physik, West Germany
	Andrew C. Millunzi, U.S. Department of Energy
	Leonard Weiss, U.S. Senate
	C. Pierre L. Zaleski, Embassy of France, Washington, D.C.
Forum Rapporteur:	*Karl Cohen*, Stanford University
5:00 p.m.	Forum Officially Adjourns

Epilogue

The Center for Theoretical Studies of the University of Miami wishes to extend their gratitude to all members of the Planning Committee, Forum Moderators, Dissertators, Annotators, and other participants for their contributions to what this energy forum may have hoped to achieve on this most critical problem of all time. At the end of this assembly, we hope we will be wiser than before!

This International Scientific Forum was supported in part by grants from Mobil Research and Development Corporation, Westinghouse Electric Corporation, Dresser Industries, Inc., Alcoa Foundation, Allied Chemical Corporation, General Electric Company, and Grumman Aerospace Corporation, and by contributions from Hamilton C. Forman, William M. Benton, and Sampson Sholes. Additional support was obtained from registration fees paid by participants.

List of Contributors

Karl Cohen, Stanford University, Stanford, California.

Jean Claude Balaceanu, French Petroleum Institute, New York.

Mike McCormack, U.S. House of Representatives, Washington.

M.E.J. O'Loughlin, Exxon Corporation, New York.

Edward Teller, Lawrence Livermore Laboratory, Livermore, California.

C. Pierre L. Zaleski, Embassy of France, Washington.

Wolfgang Sassin, International Institute for Applied System Analysis, Laxenburg, Austria.

Sam H. Schurr, Electric Power Research Institute, Palo Alto.

Richard H. Sheehan, World Bank, Washington.

Leslie Grainger, National Coal Board, London.

L.G. Hauser, Westinghouse Electric Corporation, Pittsburgh.

Narasimhan P. Kannan and **Martin M. Scholl**, the MITRE Corporation, McLean, Virginia.

Thomas H. Lee, General Electric Company, Fairfield, Connecticut.

Marcelo Alonso, Florida Institute of Technology, Melbourne, Florida.

Eduardo López-Ballori, Office of Energy, Santurce, Puerto Rico.

Bruno Coppi, Massachusetts Institute of Technology, Cambridge, Massachusetts.

John F. Holzrichter, Lawrence Livermore Laboratory, Livermore, California.

N. Douglas Pewitt, Department of Energy, Washington.

Paul J. Reardon, Princeton Plasma Physics Laboratory, Princeton, New Jersey.

Robert J. Creagan, Westinghouse Electric Corporation, Pittsburgh.

Peter Fortescue, General Atomic Company, San Diego.

William H. Hannum, OECD, Nuclear Energy Agency, Paris.

Philippe Hubert, Commission a l'Energie Atomique, Fontenay-aux-Roses, France.

W. Paskievici, University of Montreal, Montreal.

Robert S. Chen, National Academy of Sciences, Washington.

Klaus Gottstein, Max Planck Institut für Physik, Munich.

About the Editors

Behram N. Kursunoglu received the B.Sc. from the University of Edinburgh and the Ph.D. from the University of Cambridge. He was recipient of the 1972 Turkish Presidential Science Prize, is a Fellow of the American Physical Society, and is a member of Sigma Xi, Phi Kappa Phi, and Sigma Pi Sigma. Dr. Kursunoglu is a professor of physics at the University of Miami and is the director of the Center for Theoretical Studies, which he founded in 1964. He has published numerous books and articles in the fields of physics and energy.

Andrew C. Millunzi of the U.S. Department of Energy, Washington, D.C., is an editor of the proceedings of the Center for Theoretical Studies' annual conferences on energy.

Arnold Perlmutter received the A.B. from the University of Califronia at Los Angeles and the M.S. and Ph.D. from New York University. He is a member of the American Physical Society, the Federation of American Scientists, and Sigma Xi. Dr. Perlmutter is a professor of physics at the University of Miami and has been secretary of the Center for Theoretical Studies since it was established in 1964. He has numerous publications to his credit and has edited books published by the center since 1964, including the proceedings of the center's annual conferences.

Linda Scott, a graduate of the University of Miami, has been deputy secretary of the Center for Theoretical Studies since 1976, is editor of the center's Quarterly Bulletin, and has been associate editor of books published by the center, including the proceedings of the center's annual conferences.